AMERICAN
SPIDERS

To Jean Moore Gertsch
*in appreciation of your
contributions to this book*

AMERICAN SPIDERS

Second Edition

Willis J. Gertsch, Ph.D.

Curator Emeritus, Department of Insects and Spiders
American Museum of Natural History

c. 2

VNR VAN NOSTRAND REINHOLD COMPANY
NEW YORK   CINCINNATI   ATLANTA   DALLAS   SAN FRANCISCO
LONDON   TORONTO   MELBOURNE

Van Nostrand Reinhold Company Regional Offices:
New York Cincinnati Atlanta Dallas San Francisco

Van Nostrand Reinhold Company International Offices:
London Toronto Melbourne

Copyright © 1979 by Litton Educational Publishing, Inc.

Library of Congress Catalog Card Number: 78-6646
ISBN: 0-442-22649-7

Manufactured in the United States of America

Published by Van Nostrand Reinhold Company
135 West 50th Street, New York, N.Y. 10020

Published simultaneously in Canada by Van Nostrand Reinhold Ltd.

15 14 13 12 11 10 9 8 7 6 5 4 3 2 1

Library of Congress Cataloging in Publication Data

Gertsch, Willis John, 1906–
  American spiders.

  Bibliography: p.
  Includes index.
  1. Spiders—North America. 2. Arachnida—North
America. I. Title.
QL458.4.G47 1978 595'.44'097 78-6646
ISBN 0-442-22649-7

Preface

Spiders make up an important part of the animal life of the vast and diversified land that is temperate North America. That general knowledge of them is still relatively meager must be attributed to the circumstance of size, rather than to inferiority in either importance or genuine interest. By means of size and sound, birds, mammals, and other vertebrate animals monopolize the stage and divert attention. Yet only a slight change in perspective will bring into view a microcosm of tiny creatures that, hidden away in miniature jungles underfoot, live lives of unbelievable strangeness and complexity. To bring the spider microcosm into sharp focus for the general reader is the prime purpose of this book.

Our American spider heritage is a large and diversified fauna commensurate in importance with the age and size of the continent itself. Proclaiming this heritage is a large and rewarding body of literature created by students during more than 150 years of devotion. At the beginning one would mention the name of John Abbot, who, as early as 1776, began the study of spiders and other animals in the region around Savannah, Georgia. It is to be regretted that his fine paintings and accompanying notes were never published, as were those of the birds, butterflies, and moths for which he became justly famous. Thereafter, with Nicholas Marcellus Hentz, whose first contribution appeared in 1820, began a line of investigators (H. F. McCook, T. H. Montgomery, G. W. and E. G. Peckham, J. H. Comstock, J. H. Emerton, A. Petrunkevitch, and B. J. Kaston, to mention only a few) which is being continued by an ever-growing number of enthusiasts. The contribution of Americans to world araneology has been a striking one, but we have profited in even greater measure by the energy and genius of students from other lands, foreigners in language only.

Our debt to the past is a very great one, and credit for our (often presumed) deeper insight into the *Araneae* must, to a considerable extent, go to the accumulation of information by the pioneers. The facts brought together in this book are borrowed largely from a fund of information available to all araneologists and, while they reflect commendable knowledge, at the same time they reveal comparative ignorance of much in the lives of the spinning creatures.

Included as part of the record are many new facts and observations of my own that have not been mentioned before. Some innovations have been welcomed and adopted by colleagues. It is the author's hope that *American Spiders* will, in addition to its other purposes, act as a stimulus to all those eager to unearth the many details still unknown.

American Spiders appeared in 1949 and has been well received by readers from many parts of the world. Its original aims were the same as those of the present edition, to present a rounded picture of a fauna probably approaching 3,000 species, to bring attention to its unique elements, and to give as far as possible a meaningful verbal and pictorial expression of it. For such a large fauna it has been possible only to give condensed essays under the various titles of the chapters. Since a principal aim of this book was to offer matter of popular content, the author has not hesitated to bring into the writing anecdotal and correlated details that to many may be of greater interest than factual descriptive content. Further, the author has not hesitated to mention stimulating details of spiders living outside the present realm.

An important contribution of *American Spiders* has been the extensive series of colored and black and white illustrations offered in the first edition. Most of those contributing photographers, former good friends and colleagues, belong to a past generation and are now dead. For this new edition it has been possible to provide new illustrations that cover a wider range of subjects depicting the forms and handiwork of many of our spiders. For use of these many new photographs the author extends his thanks to old friends Dr. B. J. Kaston of the University of California in San Diego, and to Dr. H. K. Wallace, Professor Emeritus of the University of Florida in Gainesville. In addition, he wishes to acknowledge the many important suggestions made by Dr. Kaston for improvement of the manuscript.

W. J. GERTSCH

List of Illustrations

GRAVURE ILLUSTRATIONS

Plates 1-32 appear between pages 146-147.

Contents

1
Introducing Spiders

SPIDER PREVIEW

This book treats the spiders of the United States and Canada and is concerned almost wholly with their habits and life histories, their morphology and peculiarities, and also with their numbers and kinds. Most of us know something about spiders, but few of us are aware of the vast numbers that exist and of the great diversity in appearance and habits of the spinning creatures. Yet even a limited acquaintance soon makes it evident that spiders in many ways far outshine insects and lesser animals of much greater reputation. Thus it seems desirable that, at the very outset, a few of the striking peculiarities of the maligned spiders be enumerated.

Insects have developed wings and on them have attained the most exalted place among the arthropods. Although a wingless creature of the earth and its plant cover, the spiderling can float its threads on the breezes and fly through the air, often reaching tremendous heights and sailing for long distances. This "ballooning" of spiders has been instrumental in distributing them into new colonizing areas at a rate not possible even for insects with their wings. The rigging of ships 200 miles from the nearest land has been showered with tiny aeronauts riding on silken streamers. But, alas, disaster is the fate of those that fail to land in an environment favorable for their survival. The spider can spin a line one-millionth of an inch in thickness, but most of its single lines are ten or twenty times as thick. These silken strands possess varied properties that serve desirable ends: some strands hold struggling victims by great elasticity that allows stretching fully their length before breaking; other strands rely on sticky globules or mere tensile strength, said to be equal to fused quartz fibers. Some spider lines are of such fineness that they are impossible to duplicate; they serve admirably as markers in various surveying and laboratory instruments. An inveterate spinner during all of its life, the spider uses silk for so many different purposes that this material is the most important thing in its life, the agent that has largely determined its physical form and dominant place in nature.

1

Almost alone among the lesser creatures the spider prepares a trap to capture its prey. By their structure these traps are identified as tube webs, purse webs, sheet webs, tangled webs, and orb webs. Many of these webs are complex structures of curious form but others are innovations that accomplish their purpose by a few strategically placed lines of silk.

The orb web (Plate 12; Plates XX, XXII) has long been a symbol of the spider in the mind of man, who sees in its shimmering lightness and intricate, symmetrical design a thing of wonder and beauty. Such esteem is well merited, because the orb web is the most highly evolved of all the space webs developed by the sedentary spiders. It represents a triumph in engineering worthy of great mechanical ingenuity and learning; yet it was arrived at by lowly spiders, which even by their most ardent supporters are credited with hardly a gleam of what is called intelligence. The ingredients of almost unlimited time, of moderate compulsion to irresistible change, and the stimulus of real advantages gained have contrived to produce the two-dimensional orb web from the seemingly wasteful tangle of threads that was its origin. Instinctively and blindly the spider has followed the long path leading to its symmetrical masterpiece. The orb weavers are virtually slaves of their webs and have wagered their future on the tenuous lines. Within the limits of their circumscribed world they are supreme autocrats, but when brushed from their snares, many are clumsy, vulnerable creatures.

In accomplishing the purpose of entangling flying insects, the orb web has served the needs of the spider admirably and at remarkably small cost. Only about an hour is consumed in spinning the average orb web which, because of great damage to the lines, frequently is replaced every suitable night by the methodical spider. Yet within the orb-weaving group there are some members that have broken so completely with the past that they do not spin orb webs at all but have substituted an entirely different method of securing their prey. Instead of relying on the static but dependable round web, they spin a line, weight the end with a sticky drop of liquid silk, and hurl it at their prey much as the gaucho throws his bolas or the angler casts his line. One need not travel to the exotic tropics to find these bolas spiders; they live over most of the United States and even within large cities, seeming to prefer the trees of our formal parks. Close relatives of the bolas spiders live in Australia and Africa; one of these African cousins varies the casting procedure by spinning its weighted line around like a whirligig.

The female bolas spider (Plate 17) is a plump creature, about one-half inch long and equally wide, which sits placidly on a twig, simulating with considerable faithfulness a bud, a nut, a snail, or even a bit of bird dung. What about her mate? He is an insignificant atom no larger than the head of an ordinary pin. Precociously developed, he walks out of the egg sac fully mature, along with sisters his own size who are just beginning their life and must wait weeks and increase tremendously in size before they become sexually mature.

Spiders and their relatives are ancient animals; they were among the first creatures to leave the waters for a life on land. Some modern spiders seem to be only thinly masked replicas of creatures that were living in the Northern Hemisphere during the remote Paleozoic era, when the coal measures were still in infancy. Although more generalized than the commoner true spiders, the tarantulas and their kin have become specialists in their own fashion, and have devised new and extraordinary ways of living in a world of competition. The purse-web spiders (Plates 6, 7) live in a long silken tube closed at both ends, and have developed long fangs with which they impale insects that walk over their cylinder by biting through it. The burrowing tarantulas of the genus *Antrodiaetus* ensure privacy in their burrow home by pulling flaps of silk, which fit like folding doors, over the entrance. The trap-door spiders are accomplished burrowers and cap the opening to their chamber with a hinged trap door. One of the strangest trap-door spiders is *Cyclocosmia*, which has an abdomen hardened and rounded behind to form a plug with which it at one time was erroneously reputed to close its burrow.

Among the vagrant tarantulas (Plates 3, 4) are some that have become veritable giants far exceeding most insects in bulk and rivaling in size even the great black scorpions of Africa. In addition to a standard diet of insects and other invertebrates they are able to kill with ease and feed on frogs, toads, lizards and snakes, even including small rattlesnakes, and mice and various other small mammals due to the long, strong fangs with which they are armed. Some of the arboreal tarantulas are known to kill small birds, and they have gained one of their common names of "bird spiders" from this activity. Longer-lived than any terrestrial invertebrate are some of the great hairy tarantulas which in the southern United States do not become sexually adult until eight or nine years old and are known to live thirty years. Far more precocious are giant species of South America which mature after three or four years.

The tarantulas are vastly outnumbered by the true spiders; among these latter are the diurnal jumping spiders (Plates 26–30) which actively pursue their prey over the ground and on plants. Special tufts of adhesive hairs on the tarsi allow them great freedom of movement on smooth and precipitous surfaces and, aided by the keenest eyesight of all spiders, they emulate the carnivores in stalking their prey. Their stout bodies and legs are gaily colored and bedecked with tufts of bright hairs, pendant scales, and curious spines. With their iridescent scales gleaming like jewels in the sun, they rival the gaudiest insects. During courtship dances, the little males caper and posture before the females in such manner as to display their brilliant ornaments to best advantage.

In the corollas of many kinds of flowers hide stubby little crab spiders which, simulating the assassin bugs, seize flying insects that visit the blossoms for nectar. In keeping with this role of deception, they change from white to yellow, or vice versa, to conform with their background. Large bees and butterflies are quickly subdued by the potent venom of these crab spiders.

All spiders breathe air through orifices on the ventral side of the abdomen. In spite of their air requirements, many have adopted an amphibious life and stay underwater for periods of variable length. Some live in little waterproof chambers spun in holes in coral rock that are covered over during high tide. Most extraordinary of all is the water spider *Argyroneta* of Eurasia, which is able to swim about and live for weeks in the fresh water of streams and ponds, living mostly in a domicile that resembles a small diving bell. This spider carries air bubbles beneath the surface to its retreat, which is anchored to aquatic plants by silk lines, and keeps a supply of air imprisoned in the silken chamber. Its prey consists of small aquatic animals which it captures in the stream. Even the eggs are laid and the family hatched out underwater in the security of the nest. Among American amphibious spiders are some of the fisher and wolf spiders, which run over the surface freely and dive into its depths where they may stay for long periods. Occasionally small fish or amphibians are caught by the large fisher spiders of the genus *Dolomedes*.

The sexual characteristics of spiders are especially interesting. In both sexes the genital opening is a simple pore beneath the base of the abdomen through which emerge the spermatozoa or eggs. One would expect that during mating the male products would be transferred directly to the female by contact between these orifices or by means of an intromittent organ. Instead, the male spider has transformed the claws on the ends of the pedipalpi (the leglike appendages lying on each side of the head in both sexes) into a complicated intromittent organ, comparable to a syringe or hypodermic needle, and has modified and greatly enlarged the distal segments of the pedipalp to protect the organ and facilitate the pairing. These organs of the male, called palpi, have no internal connection with the gonads of the abdomen, so the semen must be transferred from the genital orifice to the palpi. To accomplish this the male spins a little sperm web, deposits a small globule of semen upon it, and then sucks it into the syringe in each of the palpi. The female has developed in front of the genital pore paired pouches for the storage of the semen, each unit of which is shaped to receive the corresponding palpus of the male.

Since spiders are solitary, predaceous creatures, the male runs considerable risk in approaching his usually larger mate, who may be only hungry and not ready for mating. Some males are killed because of early failure to diagnose the attitude of the female or, after being successful in their suit, of not leaving the premises before the normal predatory instincts of the female again dominate her. Various routines have been devised by different groups of spiders to gain the recognition of the female and make possible a transfer of the semen in relative safety. Once the female has been inseminated, she is able to retain the male products in the receptacles until they are used to fertilize the eggs during the egg laying.

In the bodies of spiders are found clues that give considerable insight into the racial history of the group. From lumbering ground creatures have come fleet runners on soil and vegetation and trapeze artists that hang in midair on silken lines. In the variety and strangeness of their forms spiders equal or surpass all comparable invertebrate groups. In color pattern, ornamentation, and brilliance they are on a par with any of the insects. Indeed, the vaunted brilliance of the morpho butterflies and of the birds of paradise is excelled by the iridescent variety of the jumping spiders of the tropics. Only the small size of spiders conceals their beauty and keeps them largely unknown.

Finally, it should be noted that spiders have attained their present position without benefit of so-called intelligence. Endowed with incredibly complicated instincts, these spinning creatures perform their marvels largely as automatons, and show only moderate ability to break the bonds of their behavior patterns. The baby orb weaver spins a perfect orb web soon after it leaves the egg sac, and thereafter scarcely changes it, except in size, during its whole lifetime. The mother spider encloses her eggs in a sac which, often beautifully designed, advertises the species to which she belongs, and then sometimes is on hand to defend her precious burden against assailants. Instinct plays a large role in every action of the spider and is the guiding principle throughout its life.

GENERAL ATTITUDE TOWARD SPIDERS

Spiders are seen in different lights by different peoples. Primitive men regard some spiders as bad, others as good, and most as having little importance or significance in their lives. To those that become important because of venomous or presumed dangerous character, they give special names. The *chintatlahua* of the Oaxaca Indians, the *po-ko-moo* of the Mewan tribe of California, and the *katipo* of the Maoris, all refer to similar species of the genus *Latrodectus*, which have long been notorious over much of the temperate and tropical world. Each people has a distinctive name for the brightly marked spiders known as "black widows" in North America. In addition, species resembling the virulent ones are regarded with suspicion and often endowed with the same venomous powers. This is a practical approach, learned by trial and error, and tested in time by peoples who have close contact with the lesser creatures about them. Thus it is not surprising that many beliefs of primitive peoples often have a firm foundation in fact.

In another category are some spiders that are good because their presence at certain hours, on specific occasions, in particular places, constitutes a good omen. A few are eaten with keen relish by peoples in various parts of the world. Others are seen as wonderful creatures that produce marvelous webs overnight and have magical powers.

To the American Indians the spider is a creature of mystery and power which, though capable of trickery, duplicity, and even great evil, plays a benevolent and often potent role in many of their legends. The prowess of spiders in this folklore is based largely on their great skill as spinners, and to a lesser extent on the deadliness of their bite. To the Dakotah the orb web is a symbol of the heavens; the corners of the foundation lines point in the four directions from which come the thunders, while from the spirals of the orb emanate the mystery and power of the Great Spirit. In Indian legend, spiders are venerated for spinning silken lines of great strength on which some unfortunate is able to escape from destruction. A youth, betrayed into sleep by the seduction of a woman, awakens on a precipitous cliff but lowers himself to safety on a line furnished by a spider friend. This same silken cord may also be a rope to the sky on which the dead mount to the new hunting ground, or the brave climb to wreak vengeance on the sky people. But more often it is a line from the sky to the earth on which the pursued can descend; it is on such a "sky rope" that the Algonkin maiden, fallen from grace as wife of the Morning Star, is sent back to earth.

In many interesting myths of the Pueblo Indians the main role in the Creation is assigned to the spider. According to the Sia Indians, in the beginning there was only one personage, a spider, living in a world sterile of life and lacking many material things. From each of two little packages possessed by the spider was conjured, in response to its magical singing, a woman. From the first woman thus created have descended all the Indians, and from the second all the other races of men.

Some of the virtues attributed to spiders are industry, patience, and persistence. Well known is the legend of Robert Bruce who gained new courage by watching a spider finally reach its cobweb home after many unsuccessful attempts. In a delightful Cherokee myth the little spider appears as a successful agent when all other animals fail. In the beginning the world was cold. Then fire appeared on the earth, having been placed in a hollow tree on an island by thunder and lightning. The shivering animals gazed across the waters and resolved to secure the warmth of the fire for their own purposes. After consultation, the raven was dispatched to secure the bright embers, but was unsuccessful and soon returned with the blackened feathers that it wears to this day. One by one the birds, snakes, and other animals risked a trial but all brought back only scars from the fiery furnace in the tree. Finally, the spider alone was left to brave the waters. She prepared herself by spinning a little *tusti*-bowl of her silk and then fastened it to her back. Skating across the surface of the water, she crept through the grasses to the site of the fire, caught a little ember in the *tusti*-bowl, and delivered the priceless jewel to the waiting animals. This successful venture is usually attributed to one of the amphibious wolf spiders which drags its egg sac behind it, attached to the spinnerets.

A legend of great antiquity is that of the Spider Woman of the American Southwest, who is credited with being the inventor of weaving and the teacher of all textile art to the various Indian tribes. She is an earth goddess and usually lives in a burrow deep in the soil with the Spider Man, her husband. According to Navajo legend, the art of blanket- and basket-weaving was brought to them by an unhappy Pueblo girl from Blue House, near Pueblo Bonito, who came to the hogans of the Navajos to earn her living. One day the girl wandered far from the hogan and, attracted by a thin wisp of smoke, discovered a small hole in the earth at the bottom of which was an old woman spinning a web. It was the Spider Woman (*Na'ashje'ii Asdzaa*), who quickly invited the girl to enter her house and blew up the hole until it was large enough to accommodate her guest. Befriended by the kindly Spider Woman, the girl stayed several days and learned to weave the blankets and baskets that now distinguish the Navajo. The Pueblo girl then transmitted this weaving art to her adopted people, and along with it an admonition from the Spider Woman that to forestall bad luck a hole must be left in the middle of each article. In compliance with this request, the Navajo women left a spider hole in each blanket, like the entrance to the burrow of the Spider Woman. Even to this day the spider hole may still be found in the blankets and baskets of the Navajo. The position and form of the spider hole are greatly changed and masked in deference to the wishes of persons who pay a better price for flawless examples. Needless to say, it is always present in the blankets of the old women who do not care to risk the anger of the mythical Spider Woman and the threat she made to spin silken threads in their heads.

Not always a benevolent earth goddess living in a hole in the ground, the Spider Woman is reputed to be domiciled on various buttes and crags in Navajo land. One such location is Spider Rock, a spectacular red spire towering upward hundreds of feet from the bottom of Canyon de Chelly. Navajo mothers warn unmanageable children that the omniscient Spider Woman will clamber down on her silken ladder to enswathe them and carry them back to her eerie perch. Attesting to this are the whitened bones of the victims tinting the top of the spire.

In a number of legends, spiders are pictured as villains and murderers. Thus the Winnebagos tell of the eight blind men who snared and killed people with long cords strung among the trees. Wash-Ching-Geka, the Little Hare, went among the evil creatures, incited them to quarreling, and then poisoned the meat they were cooking. They ate of the meat and were soon dead, whereupon Wash-Ching-Geka discovered that they were in reality spiders.

The duplicity of the spider is dwelt upon in the rhyme of the Spider and the Fly, and this theme also occurs in the Indian legends. Here the spider is often a rascal and excels as a trickster. The Zuni Indians tell a pleasing story of how "old tarantula" dupes a handsomely dressed youth and finally absconds with his prizes. The youth is persuaded to allow the old rascal to don his fine clothes so that he can appreciate how handsome he appears in the eyes of others. As Herbert F. Schwarz relates:

"Look at me now. How do I look?" asks the spider as he displays the garments. The youth, finding the ugliness of the wearer somewhat detrimental to the appearance of the clothes, is not greatly impressed. The spider moves off a bit, and as distance lends enchantment, or at least makes repulsiveness less obtrusive, the youth notes an improvement. Still a little farther off moves the spider, pretending that his only object is to gain the youth's approbation, but really intent on getting nearer and nearer to the burrow. At last he arrives at the entrance. "How do I look now?" asks the wily creature. "Perfectly handsome," replies the youth; but as he speaks the spider dives into the earth with the stolen finery.

Many curious beliefs are current in various parts of the United States regarding spiders, and often they are contradictory. It is generally believed that killing a spider or a harvestman will bring rain, and that many cobwebs on the grass in the morning foretell clear weather. The color of a spider is frequently of much significance in these superstitions. Black spiders are almost invariably bad, just as white spiders almost certainly signify good luck, but occasionally the colors are reversed and assume the opposite attribute. Although in some cases they are thought to be unlucky, the appearance of spiders is usually supposed to signify good luck, bringing to the observer new clothes, gifts, money, or visitors.

In contrast to the admiration expressed by the celebrated naturalist Dr. Thomas Muffet for spiders and their spinning talents, his daughter Patience, the Little Miss Muffet of our nursery rhyme, was disinclined to associate with a spider that "sat down beside her." It is suspected that her aversion was at least in part caused by the spider concoctions her father dosed her with to cure many ailments. Much of the general antipathy for spiders can be traced to teachings from parents and grandparents who early instill young children with misinformation. The Little Miss Muffet syndrome still dominates the learning program of the young and remains dominant until experience and education change the picture. In the growing fad to keep tarantulas as pets, along with a host of other small arachnids and insects, a practice sponsored by pet store owners, may well be some moderation of the negative attitude.

The spider fauna of the average lay person consists of a "black widow," a "brown spider," and a "tarantula," all considered to be bad. There seems to be a failure to understand the numbers, variety, and importance of these animals. The mostly undeserved bad reputation of a few species has been magnified beyond reason and is now attached to all of them. There is a general belief throughout the Unites States, and probably over much of Europe, that the bite of almost any spider is dangerous. Public opinion has been influenced by tall stories from far places, by sensationalism in newspaper reports, and by the natural prejudices of housewives who can be forgiven for wanting their rooms completely free of all crawling creatures. Spiders are for the most part small and, because of their nocturnal habits, rarely intrude upon our notice.

A frank dislike of spiders because of their predaceous habits would put the popular prejudice on a rational basis. The spectacle of insects being pounced upon, trussed up, crushed, and sucked dry is one that prejudices us in favor of the underdog. But we have little dislike for other creatures, such as the ladybugs, which are quite as voracious. It is doubtful that people give sufficient heed to spiders to be affected by their rapacious methods; they are labeled nasty, crawly creatures in a completely irrational manner.

2
The Place of Spiders in Nature

RELATIONSHIP TO OTHER ARTHROPODS

The vast assemblage of animals comprising the phylum *Arthropoda* includes such familiar creatures as the crabs and lobsters, centipedes, millipedes, and insects, as well as the spiders and their multitudinous kin. Indeed, three-fourths of the known animals of the world are arthropods and attest by their numbers, their variety, and their occupancy of every conceivable habitat in nature a degree of success not even closely approached by any other group of animals. Present in numbers conservatively estimated as beyond a million different species, they make up in vast populations what they concede to the vertebrates in size. Most of them are small, and because seven out of every ten kinds are insects, the average size is perhaps as small as a quarter of an inch. Indeed, it is perhaps to this small size and to superior armament in the form of a tough but light external covering, that they owe their dominance in the world.

The arthropods have their bodies encased in a stiffened outer covering, or exoskeleton, and completely lack the type of internal skeleton present in the vertebrates. The integument is made impermeable to liquids and gases and kept hard and tough by the presence of amber-colored substances called sclerotin and chitin. Between the body segments and the joints of the appendages the cuticle is not so strongly impregnated with sclerotin and remains soft and pliable, allowing movement of the legs and other articulated segments of the body. The problem of growth in size has been solved in the arthropods by their shedding the rigid outer skeleton at rather definite intervals, a process called *molting*. All the increase in size of the carapace and appendages, and often of the abdomen as well, must take place immediately following molting when the integument is still soft.

One characteristic of all the arthropods is the fact that their bodies are divided transversely into numerous well-marked rings or segments (in some cases most indications of segmentation are lost). The segments in front, which go to form the guiding center of the animal, are usually dissimilar and so greatly modified and fused that their exact limits are obscured. Thus, the head of one group is not necessarily the same as the head in another; it may be composed of more segments or carry more appendages, and the appendages of the same segment may be vastly different. From primitive appendages have been derived antennae, mouth parts, legs, swimmerets, spinnerets, and many other organs. They are used for sensory perception, feeding, running, swimming, silk spinning, mating, and other purposes. The hind portion of the animal, which is called the abdomen, is likewise not the same in all the arthropods. In the centipedes and millipedes it is a multisegmented trunk, provided with numerous jointed legs, in some instances more than 200 pairs. In the insects the abdomen completely lacks appendages except at the caudal end. In spiders the only abdominal appendages are the spinnerets.

With such marked differences in the external form of the *Arthropoda* as compared with the vertebrates, it it not surprising that the internal anatomy should also be quite distinct. The various systems for carrying on living, such as those for digestion, respiration, excretion, and reproduction, show marked differences.

In the horseshoe crabs and most of the crustaceans, the respiratory organs are external gills which aerate the blood by absorbing through their delicate walls the oxygen and other gases dissolved in the water. Whereas most of the other arthropods long ago abandoned an aquatic life, some individualists among the insects have secondarily returned to its security during parts of their life, but not before they devised new means of living. Respiration in the land arthropods is effected by means of internal aerating chambers called book lungs and tracheae, or less frequently, by breathing directly through a soft outer covering. The book lungs of the arachnids are closely packed sheets of body surface bound together like the leaves of a book, to give the maximum surface for aeration. The tracheae in the arachnids are small tubes that lead into the body and sometimes ramify to form complex systems. In the myriapods and insects, the air is conducted directly to the tissues by means of tracheae which, however, are dissimilar to those of the arachnids and develop in a different way. Although simple diffusion through the skin or into the body by means of the tracheae often is sufficient in small, inactive arthropods, in large and more active forms some sort of breathing takes place, usually through rhythmical movements of parts of the body by special muscles of the abdomen.

The type of circulatory system in the arthropods is a specialized one, seemingly highly efficient within the size limits of these creatures. Instead of closed tubes that carry the blood to every part of the body and ramify in great profusion to reach all the tissues, the system is at least in part an "open" one. The place of the veins is taken by expansive channels, or sinuses, filled with blood, in which the organs and tissues are bathed. In a large pericardial sinus lies the heart, which expands to allow the blood to enter the pumping vessel through paired lateral openings, and contracts to send the blood coursing through the arteries to all parts of the body. The blood is ordinarily a clear liquid in which are suspended numerous pale blood cells. A disadvantage of having the internal organs bathed in blood is the seriousness of accidental rupture of the outer covering. Any breaking of the body wall might prove fatal to the creature, since the blood would quickly drain from the body, but the tough exoskeleton guards against this. Injury to an appendage could also be fatal, but in many arthropods the injured member is removed by breaking it off (a process called autotomy) at a point where healing is rapidly accomplished.

The digestive system is a tube that extends from one end of the body to the other, and is often subjected to various types of elaboration by coiling and compounding to increase the amount of absorptive surface. In the spiders and their relatives, this is accomplished by extending arms in many directions from the main tube. A considerable diversity exists among the arthropods as regards the details of the digestive system, but all are alike in having a foregut and hindgut derived from the infolding ectoderm, and an expansive midgut, in which absorption is accomplished by means of the enzyme-producing epithelium.

The foregut in spiders is modified to pull in liquid food. It consists of a pharynx into which the small mouth opens, an esophagus, and a so-called sucking stomach. The former two are rather simple tubes, but the sucking stomach is an enlargement behind the esophagus supplied with powerful muscles on its four sides. When these muscles contract, they increase the size of the stomach and there results a strong sucking action that pulls the predigested food into the midgut. All the absorption occurs in the midgut, which is notable for a series of large blind sacs in the cephalothorax extending as four thick arms on each side, and voluminous glandular extensions from the main digestive tube in the abdomen. The hindgut provides a channel of egress for the fecal material, a thick, whitish liquid that is accumulated in a large bladderlike sac called the stercoral pocket, and voided through the anus.

In addition to the tubular Malpighian vessels opening into the hindgut, which serve as excretory organs, spiders have a pair of coxal glands located opposite the coxae of the first and third legs, and these discharge their products through tiny openings behind the coxae. It is believed that the coxal glands are modified nephridia, the primitive excretory organs of earthworms and other animals, and that from similar glands in other parts of the animal have developed the various silk glands and perhaps the poison glands as well.

The activities of the arthropods are governed by a nervous system quite different from that found in the vertebrates. In the simpler forms it consists of a double nerve cord lying below the alimentary tract, which is enlarged in each segment to form a center or ganglion, from which lesser nerves arise. The most anterior pair of ganglia lie above the pharynx and, joined to the pair immediately behind and below the pharynx by nerve connections, is the brain. A very considerable modification of this generalized condition is to be seen in most of the spiders which have contained within the cephalothorax all the central nervous system. The ganglia in the cephalothorax have been consolidated into a single mass around the esophagus and below the digestive system. From the dorsal brain arise the nerves for the eyes and the chelicerae, and from the lower mass large nerves go to the appendages and back into the abdomen through the narrow pedicel.

Sensation from the external environment is communicated to the central nervous system by means of structures called receptors. The most obvious receptors are the eyes which are remarkable organs in insects but by comparison very feebly developed in spiders. Also, very poorly represented in the arachnids are receptors for chemical stimuli, such as smell and taste, and perhaps the former sensation, as it is understood in vertebrates, is not even present in spiders. The receptors for touch are numerous and varied in the *Arachnida*, and it is through stimulation of these touch receptors, with attendant chemical properties, that these animals are best able to know their environment.

THE NEAR RELATIVES OF SPIDERS

The spiders and spiderlike animals belong to the class *Arachnida*, one of the major divisions of the *Arthropoda*. The nearest relatives of the arachnids are the horseshoe crabs, represented by *Limulus* in the waters along our Atlantic Coast, and the now extinct eurypterids, or sea scorpions; both of these secured oxygen through external gills. The arachnids differ at sight from most other arthropods in completely lacking visible antennae, the sensory appendages on the heads of insects and crustaceans often appropriately called "feelers." Although frequently confused with insects because of similar size and general appearance, the arachnids are not close relatives of these creatures, which have only three pairs of legs and have developed wings. All adult arachnids have four pairs of legs, except in rare instances, and they never have wings. Insects and crustaceans have at least two pairs of appendages behind the mouth, mandibles and maxillae, with sharp surfaces for cutting, chewing, or sucking plant and animal food before swallowing it. Typical arachnids lack mandibles but instead have a pair of pincerlike chelicerae which, with minor help from adjacent endites, pierce and break the prey. At the same time digestive juices flood the victim and predigest the pap to a form capable of being sucked through the small mouth.

The precursors of the arachnids were among the first animals to crawl out upon the land and full success on land demanded various innovations. Of first importance was aerial respiration: this was accomplished by covering over of the gills and transforming them into book lungs and tracheae to prevent desiccation. A prime need was the invention of a new system for safe transfer of the male sex products to the body of the female. To the sperm mass, a simple globule held together by its own viscosity, the male gave special attention and as time went on produced various kinds of protective devices, spermatophores, to support it and devised distinctive ways of passing it to the female gonopore. The various arachnid orders use different ways of accomplishing this.

Once on land the early arachnids found this habitat a suitable and challenging one for their development and almost none have returned to the water to live even a part of their lives, as have many insects. A few of the mites have invaded fresh and salt waters, where they largely live parasitically on the bodies of aquatic mammals. The early land arachnids fed on the invertebrates and insects, many of them probably amphibious, attempting to establish themselves as aerial creatures. Since then arachnids and insects have lived together as antagonists, the former mainly as predators and the latter as their prime food. Insects have responded with a remarkable adaptive radiation to occupy and become dominant in every stratum of the environment. Spider silk, first a protective blanket for the eggs, was transformed and refined into an offensive capturing device aimed at all levels of the insect environment. Some insects turned the tables and are among the worst enemies of spiders.

Important and interesting in their own right are the arachnid relatives of spiders, such as the scorpions, harvestmen, and mites, which in this book can be mentioned only briefly in passing. Each of the major groups, or orders, has developed a distinctive form and many are quite familiar to most people. Every one of the ten living orders of arachnids occurs within the limits of the United States. Only a brief resume of some of their distinctive features and habits can be mentioned here.

Order *Scorpiones*   the Scorpions
Order *Pseudoscorpiones*   the Pseudoscorpions
Order *Opiliones*   the Harvestment
Order *Acari*   the Mites
Order *Solifugae*   the Solpugids
Order *Ricinulei*   the Ricinuleids
Order *Uropygi*   the Tailed Whip Scorpions
Order *Schizomida*   the Schizomids
Order *Amblypygi*   the Tailless Whip Scorpions
Order *Palpigradi*   the Micro-Whip Scorpions
Order *Araneae*   the Spiders

Scorpions. Among the oldest and most generalized members of the land arachnids, the scorpions are often called living fossils because they have changed so little since the Silurian period, about 400,000,000 years ago. Among the oldest is a remarkable species from fossil beds at Waterville, New York, given the name *Proscorpio osborni* and long presumed to be one of the first animals to adjust itself to land life in North America. This ancient creature, now known to be aquatic, lived in the same waters as some of the eurypterids, the giant water scorpions, a now extinct sister group which at some time in its long history may have given rise to the true scorpions. Such aquatic scorpions had a single large claw on their tarsi, had large compound eyes on each side of the carapace, and breathed by external gills. Not until much later, in the Upper Devonian period, about 345,000,000 years ago, did scorpions learn to live on land. The scorpion spermatophore develops in the body of the male and features an injecting mechanism that propels the sperm mass into the gonopore of the female. The mating dances described so vividly by Fabre are merely the efforts of the male scorpions to maneuver the female to a suitable smooth surface to make possible transfer of the sperm mass.

About 800 species of scorpions live in tropical and warm temperature parts of the world; more than fifty live in the United States. The most obvious characteristic of the scorpion is the elongated, taillike projection of the abdomen which bears a poisonous sting at the tip. In life the tail is curved over the back, and the spinelike sting is directed forward, always in position to attack its prey. The sting is generally used in conjunction with the large pedipalpi, which are developed as pincers to grasp and hold the victim. The sting of most scorpions is painful and the venom is capable of causing mild to severe local reactions. The venom of a few species, notably those of North Africa, Brazil, western Mexico, and Arizona, carry nerve toxins that paralyze the respiratory muscles and cause cardiac failure in warm-blooded animals and man. Several dangerous species of *Centruroides* in Mexico are responsible for the deaths of many children each year. A similar species in Arizona, *Centruroides sculpturatus*, is less venomous but still considered far more dangerous than the black widow spider.

Scorpions produce living young that soon after birth mount the back of the mother and stay there until after their first molt, usually for a week or more. During this time they do not feed, but rely for sustenance upon the food stored in their bodies. The story that these weak creatures feed upon the body juices of their mother is a figment of some fertile imagination. Another fable is the belief that scorpions commit suicide by stinging themselves when they are hopelessly cornered or surrounded by a ring of fire.

Pseudoscorpions. The pseudoscorpions are so named because of their superficial resemblance to true scorpions. They have the same enlarged pedipalpi terminating in pinching chelae, but the segmented abdomen is broadly rounded behind and is without the trace of whip or sting. Of the more than 2,000 species in the world, the largest ones are scarcely more than one-fourth inch in length and most of the others are much smaller. These creatures, usually brownish or amber in color, live under stones, in moss or other debris on the ground, under the bark of trees, in the nests of ants, bees, and termites, and sometimes in the dwellings of man. Many are found only in caves and these are of pale yellow or whitish coloration. One of the better known species, the large cosmopolitan *Chelifer cancroides*, lives in houses and shelters of man all over the world.

The food of pseudoscorpions is believed to consist mostly of small insects and their larvae, more rarely mites; these are grasped with their claws, torn with the strong chelicerae, and perhaps even anesthetized by venom from tiny glands in the chelicerae. Along with spiders and some of the mites, the pseudoscorpions share the ability to produce a kind of silk. It comes from glands that are probably homologous with those that in spiders produce the venom to subdue prey. During periods when they are relatively helpless, such as when the female is distended with eggs or when molting, they spin silk copiously and enclose themselves in wonderfully constructed retreats. The pseudoscorpions have poor vision, and frequently the eyes are lacking altogether. Many blind species live in caves and in deep ground detritus in outside habitats. The numerous sensory hairs on the pedipalpi and on other parts of the body take the place of eyes. Some pseudoscorpions practice a curious, perhaps protective device called phoresy: they attach themselves to the bodies of flying insects such as flies, bees, and beetles and are thus carried quickly from one locality to another, perhaps away from an overcrowded environment.

At the time of reproduction the male pseudoscorpion produces and places on the substratum a distinctive spermatophore, consisting of a thin stalk on the tip of which is placed a sperm globule. This is picked up by the female in various ways: in some species the spermatophores are placed randomly in areas inhabited by females and she is able to find them by chemotactic means; in other cases the male confronts the female, practices an elaborate courtship ritual to entice her to pick up the sperm ball, or even forcibly opens her gonopore to engulf it.

Harvestmen. Familiar to most people because of the great length of their legs, the harvestmen (phalangids or daddy-longlegs are also acceptable names for this group of about 3,200 species) scarcely need introduction. Though often confused with spiders to which they have a certain resemblance, they can always be distinguished by their body which has the cephalothorax and abdomen broadly joined to form a single suboval unit. In this respect they are similar to the mites but differ from them in that they usually have an abdominal portion that is made up of well-marked segments. By way of confusion, it can be reported that there are mitelike harvestmen and harvestmenlike mites, classifications that emphasize the close relationship of these arachnid orders.

Most harvestmen found in the temperate zone are active animals that run rapidly on stiltlike legs; these are sometimes thirty or more times as long as the body. When in danger of being caught or actually seized by a predator, they shed some of their legs by breaking them off between the coxa and trochanter and then run away while the detached appendages continue to quiver in front of a probably confounded predator. Autotomy is also a protective device of spiders and some other arachnids. Some phlegmatic harvestmen feign death and get away with this deception. The presence of scent glands in some species may make some species unpalatable. Harvestmen often congregate in large numbers on vegetation or on the trunks of trees and, are especially noticeable during the harvesting season, a fact that has inspired the common name. The harvestmen feed on a variety of living and dead invertebrates, mostly springtails, collembolans, and other small insects, but also on mites and even snails.

The long-legged harvestmen are partially replaced in the warmer parts of the United States and in the tropics by shorter-legged species that tend to be less active and frequently are quite sluggish. Many of these are bizarre animals with beautifully sculptured bodies often set with strangly shaped spines and processes. Many strange species live in caves in our southern states and over the warmer parts of the world; some of these are blind troglobites that occur in a single cave or cave system and can now live only in such restrictive habitats.

The sexual biology of the harvestmen is most interesting because of the development in the male of an elongated copulatory organ, the penis. There seems to be no formal courtship or prenuptial ceremonies and the male simply climbs on the body of the female and introduces the spermatozoa into the gonopore of the female. Most arachnids lack such a copulatory organ and use an indirect method to accomplish this by means of a spermatophore and various accessory appendages. This simple, safer way of transfer of the sex products has many advantages, but some students believe that the penis is a secondary development that came after use of a conventional spermatophore stage. The female buries her eggs by means of a long, eversible ovipositer into the soil, under the bark of trees, or into the stems of plants.

Mites. Mites far surpass the other arachnids in number of individuals and undoubtedly also in number of species; the more than 20,000 species so far known probably represent only a small percentage of the total. Most are minute reddish creatures with unsegmented, ovoid bodies fused into a single piece. The tiniest mites are wormlike and suck plant juices, thereby causing galls, spots, and blemishes on the foliage of trees and plants. Other mites live in the tracheal tubes of bees and in the hair follicles of mammals, including man. Some of the more gaily colored species have taken to living in water and swim with the aid of long hairs on their legs. The free living forms abound in the ground and other detritus, where they prey on tiny animals or eat decaying animal or vegetable matter. About half the mites live parasitically on the bodies of animals all or a part of their lives. Mites are of great economic importance in our agriculture, are responsible for transmitting serious diseases of man and his animals, and are for the most part regarded as obnoxious creatures.

Mites hatch from eggs as six-legged "larvae," an unusual physical stage for which we still have no good explanation. After a period of feeding the larvae change into eight-legged nymphs, which then undergo one or more nymphal feeding stages before becoming sexually mature adults.

Most pestiferous of all are the larvae of the harvest mites, known to Americans as red bugs or chiggers; these mites attach themselves to the skin and cause violent itching and irritation. Some red bugs transmit rickettsial organisms which cause tsutsugamushi disease or scrub typhus, frequently fatal to man. The nymphs and adults of the red bugs are innocuous creatures content to live on vegetable matter.

Largest of all the mites are the ticks whose leathery bodies are capable of becoming greatly distended with blood, to nearly an inch long in some females. Following engorgement, which is accomplished by forcing the beaklike mouth parts deep into the skin of the host, the mature females fall to the ground and lay several thousand eggs. From these eggs hatch six-legged larvae called "seed ticks" which climb on the body of a new host when the opportunity arises. Some ticks use the same host during all their feeding (the brown dog tick is such a pest) but others require two or even three different kinds or sizes of hosts in order to complete their life cycle. Many ticks attack man and are a great source of annoyance because of their irritating bite. Among tick-borne diseases are Texas fever of cattle and Rocky Mountain spotted fever, a serious illness of man carried by *Dermacentor* ticks in many of our northern states.

The sexual biology of the mites is more complicated than that of other arachnids in that several means are used to pass the spermatozoa to the female. Many use an indirect method with production of a spermatophore; this is offered to the female on a stalked spermatophore or forced upon her by various appendages.

Solpugids. The curious arachnids known as solpugids (Plate I) or wind scorpions are ancient animals possessing many primitive features and are first known to have appeared in the Carboniferous period. The relatively soft suboval abdomen has ten ringlike segments and the pedipalps are simple leglike appendages. The very large, powerful chelicerae are not matched in relative size by any other arachnid group. While feeding, the two giant chelicerae work with a sawing motion, holding back with one unit while the other is driven in deeper, to form a grinding mill that crushes and breaks their prey. It is believed that solpugids, like other arachnids, take only liquid or finely particulate food into their small mouths. Insects of many kinds and sizes make up their food and these include beetles, moths, and in the Southwest many favor termites. Solpugids are active runners and, while foraging mostly at night in random fashion over the ground or other substrata, they find their prey mainly by touch and vibration. Some species climb actively on shrubs and trees and many have tarsal pads that even allow them to climb up on glass and smooth surfaces.

The solpugids have devised their own distinctive method of passing the sperm droplet to the female and, since no formal spermatophore is used, it may represent an early stage of the process. The male confronts the female, often seizing her with his chelicerae, and if she is ready for mating she responds by becoming quiescent. In many species the male voids a sperm droplet on the substratum, then picks it up by his chelicerae and transfers it to the female gonopore by means of these appendages. An even safer innovation is practiced by our American species of *Eremobates:* the male turns the female over, emits a droplet of seminal fluid from his genital opening directly onto the female gonopore, and then proceeds to ensure entrance of the fluid into the storage area by tamping it in with his chelicerae.

The solpugids, already known by more than 800 species, abound in Africa and the Near East where many large and exotic types occur; perhaps the largest one is *Galeodes caspius* of North Africa which has a body length which sometimes exceeds two and one-half inches. By comparison most of the more than 100 American species mainly of our Southwest are pygmies, being mostly about an inch long but with a few nearly double that size. Their long legs covered with reddish hairs give these belligerent animals a formidable appearance. The solpugids possess no poison glands, cannot effectively use their chelicerae on large objects, and need not be feared by man.

Ricinuleids. The curious, enigmatic arachnids of this group, first known from fossils in the Carboniferous period of the Northern Hemisphere, now live in tropical West Africa and in the Americas north to southern Texas. Once considered to be the rarest of all arthropods, they are now known by more than thirty species and some of these by numerous individuals. The dark brown ricinuleids resemble ticks superficially in general appearance and further simulate the sluggish, deliberate movements of the latter. Various structural peculiarities set the ricinuleids apart from all other living arachnids. No true eyes are present but vague pale spots on each side of the carapace may be vestiges of obsolete clustered ocelli. Appended to the front edge of the carapace is a convex hood, the cucullus, which fits down tightly over the small chelicerae. The cephalothorax is narrowly joined to the abdomen by a pedicel, but this is hidden from view by expansions of the base of the abdomen, which fits very closely with the cephalothorax, the juncture forming a tight coupling device. The living animal is able to disengage the carapace from the abdomen so that the genital orifice is exposed, an action necessary during egg-laying and mating. The third leg of the male is provided with a complicated copulatory apparatus presumed to aid during the mating. The structure of this unique organ is suggestive of that of the spider palpus and perhaps it receives the sperm directly from the male gonopore and then deposits it in the seminal receptacles of the female.

A rare species, *Cryptocellus dorotheae*, occurs in southern Texas where it was found beneath blocks of deeply embedded concrete. Several related species have been found in Mexican caves where they feed on small flies and other insects found on bat guano; some of these species seem to be limited to single cave systems.

Tailed Whip Scorpions. The uropygids or tailed whip scorpions (often placed in the order *Pedipalpi* along with some of the groups that follow) are reddish brown to blackish animals somewhat resembling scorpions but lacking a caudal sting. At the end of their flattened bodies they have a slender whiplike flagellum, or tail, which accounts for their common name. In this group the carapace is longer than broad, the pedipalpi are stout grabbing appendages, and the very thin first pair of legs form small lashes. Mostly nocturnal in habit, the uropygids spend the day in crevices in trees, underground objects of various kinds, and in burrows in the soil. In the Southwest they do not become active until the rainy season and then wander about on the ground and sometimes into houses. Although greatly feared by uninformed people, the whip scorpions are without poison glands and capable only of pinching with their clumsy pedipalps. These appendages serve admirably, along with the chelicerae, to crush and fragment beetles, crickets and other insects, and invertebrates that make up their food. These creatures have abdominal glands that secrete a liquid consisting largely of acetic and caprylic acids which is ejected through openings at the base of the flagellum. This vinegary spray, accurately aimed at lizards, mice, and other small enemies, is an effective protection against such adversaries and unpleasant even to man.

About seventy-five uropygids are known and most of them live in the humid areas of southern Asia and the East Indies. A single well-known species occurs in our region, *Mastigoproctus giganteus* (Plate I), the giant of the order with a body often three inches long. Many common names have been given to this arachnid: in Florida it is known as mule killer, grampus, or vinegaroon, mostly the latter; in the Southwest it is known as the vinegaroon or vinagrillo.

The mating behavior of the *Uropygi* resembles that of the scorpions. There are brief courtship movements by both sexes after which the male deposits a spermatophore on the substratum. Then the female moves over it and engulfs the sperm mass in her gonopore.

Schizomids. The schizomids are small amber-colored arachnids similar to the uropygids and are included by some in that order; they are kept separate here because of various discordant features. The prosoma, analogous to the carapace, is divided into a smooth propeltidium lacking eyes and followed closely behind by two free tergites. The pedipalpi are leglike with small raptorial tips, the first legs are thin and unmodified, and the long abdomen has a short flagellum at its tip. The flagellum of the female is a short, three-segmented tail but that of the male is modified to a suboval, triangular, or otherwise shaped element. The schizomids are secretive, nocturnal creatures that live under stones, logs, or other ground objects, and under surface ground detritus.

Their sexual biology is of especial interest because the flagellum of the male is put to use by the female who grasps it in her chelicerae. After the male attaches a spermatophore to the substratum, he pulls her forward to a position where she can engulf the sperm mass in her gonopore.

About seventy species of schizomids are known from Africa, Asia, and the Americas. Our few native species occur in California, Arizona, Texas, and Florida, but many more have been described from Mexico recently, where they are often found in caves. *Schizomus floridanus* of southern Florida and the Keys is known only from many female specimens and is presumed to be parthenogenetic. Our largest species, *Trithyreus pentapeltis*, which is about three-eighths of an inch in length, is common in the palm canyons near Palm Springs, California.

Tailless Whip Scorpions. The amblypygids or tailless whip scorpions (formerly also held within the order *Pedipalpi*) are reddish brown arachnids somewhat resembling the uropygids but completely lacking a flagellate whip at the end of the abdomen. The carapace is broader than long, the pedipalpi are great spinose, raptorial appendages, and the first legs are modified into very long, lashlike whips, the tips of which are flexible. These animals live in fissures in and under rocks, under the bark of trees and other dark, sheltered places. They frequently congregate in great numbers in humid caves and some of them enter houses. At night they move over walls and rock faces with their long front legs testing the terrain and they escape with great speed when disturbed. The raptorial pedipalpi, armed with rows of sharp spines, are efficient weapons that impale and crush the insect prey and then transfer it to the chelicerae to further prepare it for pre-digestion by stomach liquids. The amblypygids lack poison glands and are capable only of pinching with their pedipalps.

Mating in the amblypygids is a more protracted process than it is with the uropygids. After some stroking of the female with his whiplike front legs, the male deposits an empty spermatophore stalk on the substratum and then after more courtship play deposits two masses of sperm side by side on the top of the stalk. After the male retreats the female moves in to pick up the sperm masses with her gonopore.

There are about seventy known species of amblypygids that live mostly in the tropical situations of Africa, Asia, and the Americas. Two species of *Paraphrynus* (a generic name now used in preference to *Tarantula*) live in southern Florida and another uncommon species is found in southern Arizona. The largest species of the order is *Acanthophrynus coronatus* which has a body only about two inches long but has long appendages that can cover the area of a large dinner plate.

Micro-Whip Scorpions. As their common name suggests, these tiny whitish arachnids resemble the schizomids but they are even more generalized in their structure. The largest species so far found are only one-tenth of an inch in body length and half of that is made up of the slender tail. The palpigrades have no eyes and their mouth parts are extremely simple. All the appendages are leglike and none have become specialized for grasping, cutting the prey, or otherwise aiding during feeding. The palpigrades are considered by some to be among the most generalized of all the arachnids.

These minute arachnids are found in Texas, California, and in other warm parts of the world. They live deep in soil detritus or under deeply buried stones or logs.

THE STRUCTURE OF SPIDERS

To understand more fully the accomplishments and limitations of spiders, it is essential to have a brief resume of their most obvious physical features (Fig. 1). In common with most *Arachnida*, they have the body divided into two principal

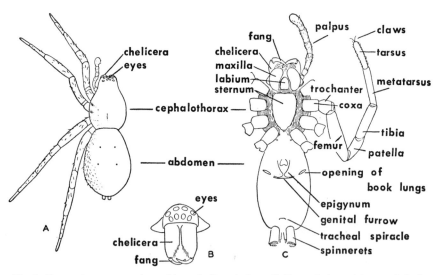

Fig. 1. External anatomy of a spider. A. Dorsal view. B. Frontal view of face and chelicerae. C. Ventral view, most legs omitted.

regions, the cephalothorax and the abdomen, and each of the sections is provided with certain types of appendages. In spiders the division between these two units is a very narrow pedicel, whereas in such relatives as the scorpions, ticks, and mites the waist is thick. From the several narrow-waisted arachnids the spiders are immediately differentiated by their possession of ventral spinning organs, or spinnerets, on an abdomen that is unsegmented except in rare instances. Furthermore, it can be noted that the males of all spiders have a complicated copulatory organ on the end of the pedipalp, a structure never found in this position in the other arachnids.

Cephalothorax. As the name implies, the cephalothorax represents those segments commonly called head and thorax, but they are intimately fused into a single piece. It must be remembered that several distinct segments have formed this region; their number is indicated by the number of pairs of appendages (in spiders only six) and by vague indications of others. The dorsal part of the cephalothorax is provided with a hardened shield or carapace, ordinarily convex and bearing the eyes at the front end. The head portion is usually more elevated, and may be strongly marked off by a V-shaped groove. On the rounded, flatter thoracic portion are usually evident a median or cervical groove and radiating depressions that mark the internal attachments of the muscles of the stomach and of the legs.

The cephalothorax is subject to considerable variation in shape and armature. In long spiders it is usually long, and in short species it may be wider than it is long. Various spines, humps, and prominences of many kinds often surmount it; frequently, some of the eyes sit on weirdly designed elevations. In the dwarf spiders the carapace of certain males is grotesquely formed, and some have deep pits into which the chelicerae of the females are fitted during copulation. In many instances the reason for the presence of such specialized innovations is not clear.

On the front of the head are the eyes which are simple and resemble the ocelli of insects. Most spiders have eight eyes, apparently the original number, but various lines have lost some, so that there are in existence six-eyed, four-eyed, and two-eyed spiders. In one tiny spider from the jungle floor of Panama only a single median eye is present, representing the fusion of a single original pair. Some of the cave spiders and others that live in dark situations have completely lost their eyes, or retain only vestiges. The size and position of the eyes vary considerably. Some of the hunting spiders have large eyes and relatively keen vision, this being one of the necessities for their foraging activities. In many, a tapetum, which causes the eyes to shine in the dark when struck by light rays, contributes to the efficiency of this night vision. Most spiders, however, are shortsighted animals that rely on their sense of touch, which they have sharpened at the expense of their eyes.

Immediately below the carapace on the ventral surface of the cephalothorax is a median plate, frequently heart-shaped, called the sternum. In front of the sternum is the much smaller lower lip, or labium, which forms the floor of the mouth. Around each side of the sternum are the coxae of the legs and the pedipalpi, which fit snugly against the sternum and lie in the space between it and the carapace. The coxa of the pedipalp in most spiders is fitted with an enlarged, sharp plate, the maxilla or endite, which aids in the breaking of the prey.

Directly beneath the cephalothorax at the front end are located the two chelicerae, or jaws, which are the offensive weapons of the spider. It is believed that the chelicerae are derived from the same pair of primitive appendages that became the second antennae in the crustaceans, and this fact illustrates the quite distinct use to which the same generalized appendages are put by different creatures. Each chelicera is composed of two segments, a basal segment, which is stout and ordinarily margined by a toothed groove at the distal end, and a shorter, movable fang, which lies in the groove when at rest. The sharp fang is the part that is thrust into the prey. Near its end is a tiny opening through which venom flows into the wound. The poison glands, said to be present in all but two small groups of spiders, are associated with the chelicerae, sometimes being entirely contained within the basal segment, but in most true spiders extending farther back into the head as more or less voluminous pouches.

All spiders are predaceous and most often subsist on the body juices of living animals; on some occasions they may accept liquid foods or be duped into accepting dead insects. The bulk of their food is living insects which are subdued by the spider's venom. Their method of feeding is a most unusual one. The strong chelicerae, with aid from the sharp edges of the endites, are used to crush and break the fresh body of the prey, which at the same time is bathed with quantities of digestive fluid from the maxillary glands. The softer parts of the prey are broken down and predigested to a liquid state, and this liquid is sucked into the stomach by means of powerful muscles. As the prey is rolled and chewed, it gradually becomes smaller and smaller until only a small ball of indigestible material remains. This is finally cast aside, or in some instances, is hung up on the egg sac or in some section of the web. In some hard-bodied insects the juices are sucked through holes made by the chelicerae, and the shell of the drained insect is then discarded. Some spiders require several hours of nearly continuous effort to digest completely an ordinary fly. It is doubtful that spiders ever actually imbibe solid food material through the small mouth, and it is certain that snake, bird, and mammal prey must first be reduced to a liquid pap by the powerful digestive juices.

The remaining appendages of the cephalothorax are the pair of pedipalpi and the four pairs of walking legs. The former are situated on each side of the mouth and resemble the legs closely except for size and for the lack of the metatarsal segment. In the female, the pedipalp is a simple appendage terminated ordinarily with a single tarsal claw, but in the male the distal end is the seat of the special copulatory organ of that sex. The role of the palpi in mating will be mentioned later.

Four pairs of legs are always present, as in typical arachnids. Each leg consists of seven segments which, beginning with the one that fits snugly into the sternal space, are named coxa, trochanter, femur, patella, tibia, metatarsus, and tarsus. At the end of the tarsus are to be found two or three claws. The legs vary tremendously in length among different spiders, some of them being long, fine stilts on which the spider hangs, and others stubby props.

With so many walking appendages, the means of synchronizing all of them is of some interest. In order to take a step, the spider moves the first and third legs of one side in conjunction with the second and fourth legs of the other side of the body. The remaining legs of both sides go into action while the other series is at rest, and thus the creature advances step by step.

The appendages and other parts of the body are usually covered with hairs and spines of different kinds. Some of these lie flat against the integument and serve as a covering blanket. Others are heavier or longer or more erect, and are used in many ways by the spider to perform important functions during the spinning of silk, for the preening of the body, preceding and during the mating, and as aids in capturing and holding the prey. Many of these setae are extremely sensitive to touch and vibration and some may be receptors for various chemical stimuli. By means of its sensory hairs the spider has a keen knowledge of its surroundings.

Abdomen. The juncture between the cephalothorax and the abdomen is made by a narrow waist or pedicel, which represents the first true abdominal segment. In the antlike spiders the pedicel is visible from above as a small tubular connection armed above and below by hard plates, but in most other spiders it is not evident, its presence being largely masked by the overhanging abdomen. Through the tiny channel of the pedicel must pass the several structures essential to maintenance of life in both body parts: the ventral nerve cord, a large artery, part of the midgut and, frequently, numerous tiny tracheal tubes.

Ordinarily, the abdomen is a saclike structure without visible segmentation and, though covered by a sclerotized cuticle, is usually much softer than the cephalothorax. In the primitive liphistiids and their relatives, the dorsum of the abdomen is armed with a series of hard transverse plates, or tergites, each set with erect black spines. In a few of the primitive true spiders there are evidences of dorsal segmentation, especially in the spiderlings, but in some well-known cases this segmentation has been acquired secondarily.

The abdomen frequently exhibits on its upper surface a series of small, rounded depressions that mark the internal attachment of muscles. Often brightly painted, and variegated with contrasting colors, the abdomen in many groups of spiders is accorded more than its share of elegance and elaboration. In some spiders the dorsum is covered completely or in part by a hard plate, and in others it is armed with curious spines and processes, some of them of great length. The reasons for the possession of such structures are no more apparent than are the reasons for those on the cephalothorax. Perhaps, because of its many sharp projections, this armor discourages birds from attacking the spider. In some of our sedentary spiders the abdomen is drawn out into a long tail, giving the creature a wormlike appearance.

The underside of the abdomen is much like the upper in many spiders, and rarely bears conspicuous prominences. Near the base are usually to be seen the two slitlike openings to the book lungs, and between them the similar genital opening. The copulatory organ of the mature female, the epigynum, is located just in front of the genital opening or hidden inside the lip. Farther back may be present a second pair of book lungs, a pair of tracheal spiracles, or, near the spinnerets, a single median spiracle. At the tip of the abdomen is the anal tubercle, or postabdomen, which has the anal opening at its tip.

Both book lungs and tracheae are found in spiders. The opening to the former is a rather conspicuous transverse spiracle, and the area of the lung itself is usually evident externally as a paler patch. In all the tarantulas and their allies, and in one small family of true spiders, two pairs of book lungs are present, the front pair near the base of the abdomen at each side of the genital pore, and the hind pair much farther back near the center of the abdomen. The presence of four lungs is usually considered to be a primitive condition inasmuch as higher spiders have the posterior pair changed into tracheal tubes. The tracheae always replace the book lungs when the latter are lost, and are not new creations at all but only modified and expanded book lungs without the leaves that ramify beyond the original space limits. In most of the true spiders, there is a single tracheal spiracle just in front of the spinnerets. In a few tiny spiders all the book lungs have been replaced by tracheal tubes.

Because in the higher spiders the book lungs have been replaced, at least in part, by tracheae, it can perhaps be concluded that these latter are more efficient respiratory organs. The true spiders are more vigorous creatures of much smaller average size than the four-lunged spider, and perhaps require superior respiratory as well as other equipment to maintain their place in the extremely diversified habitats they occupy.

The spinning organs of spiders are the spinnerets, fingerlike appendages usually located near the end of the abdomen on the lower surface. They are believed to have been derived from two-branched abdominal appendages of ancient spiders, or their precursors, which were originally put to some other use than that of spinning, perhaps being used as swimming or ambulatory organs. Associated with each of these appendages was a coxal gland in the abdomen that voided its excretory products through a pore on some part of the appendage. From the two pairs of two-branched appendages of the third and fourth abdominal segments have come the four pairs of spinnerets of contemporary spiders. Their development, modification, and elaboration have gone hand in hand with a metamorphosis of the lowly coxal glands into a series of abdominal receptacles for production and storage of distinct types of liquid silk. Originally an excretory product, silk has been put to varied and distinct uses, and it has largely charted the course spiders have followed through their racial history.

The spinnerets were originally located much nearer the base of the abdomen than their position in most modern spiders now indicates, and there was a considerable open space between them and the anal tubercle. The trend has been toward reduction of the number of abdominal segments, and the simplification of the systems inside the abdomen, as well as the segmentation of the outer integument. As the posterior segments became superfluous and were lost or incorporated into the anal tubercle, the relative position of the spinnerets also changed. Ancestral spiders had a long interval of segmented abdomen between the spinnerets and the anal tubercle. In *Liphistius* this space has been greatly reduced by partial reduction of the size of the segments. In *Atypus* and *Antrodiaetus* the space interval has been still further reduced, and in almost all other spiders the spinnerets are immediately adjacent to the anal tubercle.

Only in the most primitive spiders are eight spinnerets still present as fingerlike projections. The liphistiid spiders have retained all of the projections, but both the anterior and posterior median spinnerets are greatly reduced in size, and figure little or not at all in their spinning. In *Heptathela* only six spinnerets are present; the so-called seventh one is the fused remnant of the posterior median pair, a "colulus" in an advanced stage of obsolescence. In the other mygalomorph spiders, the anterior median pair has been lost, and in only a few are there vestiges of the anterior lateral spinnerets. Thus, the four spinnerets of the tarantulas and most of their allies represent the single, small posterior median pair and the longer, posterior, lateral, segmented pair. The loss of the spinning function seemingly has preceded the degeneration and obliteration of the spinning organs.

Most true spiders have retained the eight spinnerets in one form or another, and in only a few instances have they reduced the number below three pairs. In all the cribellate spiders the anterior median pair is still retained as the cribellum, a flat spinning field which is used in conjunction with a comb of hairs on the fourth metatarsus, the calamistruem, to produce characteristic threads. The cribellum probably existed before the anterior median spinnerets had lost their spinning function, and became greatly changed and important because of its special function. Whether the cribellum is a new development from the ancient anterior median spinnerets, or represents the ancestral condition of all true spiders, is still a debatable issue. At any rate, in almost all higher spiders a vestige of variable size evidences the former presence of an anterior median pair of spinnerets. In some it is a fingerlike colulus; in others, a pair of flat plates, connate plates, or a single sclerotized plate, all set with covering hairs; and in still others, a tiny point or blister bearing one or two erect setae. In some groups of true spiders the hind spinnerets are greatly reduced in size and become obsolete to a considerable extent, but their former location is marked by some sort of vestige.

The ordinary true spider has three pairs of well-developed spinnerets set close together in a single group. The anterior pair is two-segmented, and the apical segment is bountifully supplied with many spools and a fewer number of spigots on the spinning field. The posterior pair is likewise segmented, most commonly with two, but frequently with three or more segments, and is also well supplied with spinning equipment. Between the latter are the median spinnerets, each of a single segment, and ordinarily less well provided with spinning openings.

The spinnerets of the sedentary orb weavers and comb-footed spiders, which are the finest spinners, are relatively short with small apical segments, and they are set closely together in a small field. In many other spiders whose spinning is less noted, the spinnerets are sometimes long and conspicuous, frequently many-segmented, and arranged in different ways.

3
The Life of the Spider

BALLOONING

Much of the adventure in the life of the spider is crowded into the first few days
of freedom when the young spiderlings, having just broken through the egg sac,
strike out for themselves in a world completely new to them. It is spring and the
warmth of the sun has changed the inertia of earlier life in the egg sac to one of
intense activity. Hundreds of brothers and sisters, still closely packed together
and indistinguishable one from the other, move about within the narrow confines. Finally the actions of a few vigorous leaders result in the opening of a
small aperture at some point in the sac, and a little body squeezes through it to
greet the open air. One by one the tiny creatures emerge through the round
opening, until the sac is covered with them. They do not tarry long (Plate XVII)
but climb all over the dried leaves and stems on which the sac is anchored, stringing their threads as they go. Soon we see a tangle of webbing, strung on every
available support, crisscrossing in all directions, and invading space like a living
thing. Many of the spiderlings hang motionless once they have gained their particular station, but others press on with undiminished activity. Up and up they
climb, to the tips of the tall grass stems and the summits of the leafless shrubs
which mark the meadow site of the egg sac. Straight toward the sun they climb
until they can climb no higher, impelled by a strange urge to throw silken
threads out upon the soft breezes.

This is the urge toward ballooning, one of the most extraordinary accomplishments of the spider.

Once the spiderling has reached the summit of the nearest promontory, a
weed, a spike of grass, or a fence rail, it turns its face in the direction of the
wind, extends its legs to their fullest, and tilts its abdomen upward. The threads
from the spinnerets (Plate XXX) are seized and drawn out by the air currents.
Although the dragline threads are often used, those from several spinnerets may
stream out in long filaments. When the pull on the threads is sufficiently strong
to support the weight of the aeronaut, it lets go of the substratum and is pulled

into the air. Spiderlings balloon in different ways, and some of them when afloat climb on their threads like little acrobats, pulling in and winding up or streaming out more filaments, and in this way exercising some control of the ship they are flying.

Not the exclusive habit of a single species, as was once supposed, or limited to part of any season, ballooning goes on during much of the year and is easy to observe. In the spring and during the fall months, when immense quantities hatch from the egg, emerge from their egg sacs and fly, the ballooning spiders by their very numbers force themselves upon our attention. Small spiders can be inspired to take off if one blows steadily against them; they tilt up their abdomens, assume a ludicrous pose, and then bound into the air. Because they are so tiny and weigh an insignificant amount, spiderlings are sometimes at the mercy of the air currents and are lifted into the air when they least expect it. Even larger species, caught while dropping on their threads, are blown some distance. The small aeronauts seem to float on streamers only a yard or two in length, but the lines may actually be several times as long. In the days of Aristotle, it was commonly believed that the spider could shoot out its silk as the porcupine was once reputed to shoot its quills. We know now that the spider must depend on breezes to pull the threads from its spinnerets and to bear it aloft after the volume of silk is great enough to support its weight on the air currents.

How far do spiders fly on their silken filaments? Darwin recorded the arrival on the *Beagle* of "vast numbers of a small spider, about one tenth inch in length, and of a dusky red color," when the ship was sixty miles from the coast of South America. He watched them and observed that the slightest breeze was sufficient to prompt them to sail rapidly away, after letting out new lines to catch the wind. Even greater distances have been covered by these tiny aeronauts, which have been known to alight upon the rigging of ships more than 200 miles from the nearest land. Because they move upward and forward at a substantial pace, and because of their tiny size, the spiderlings are quickly lost to sight. The average distance they span can only be conjectured. The spider may be dropped to earth near the site of its departure, but it can fly again and again to accumulate a substantial dispersal distance.

Most ballooning goes on at reasonable heights, probably less than 200 feet as was noted by McCook; but sometimes powerful air currents carry the creatures to greater heights. During an aerial survey in Louisiana, B. R. Coad found spiders and mites well represented in samples of aerial fauna at 10,000 feet, and they were even more frequent in catches of from 20 feet up to 5,000 feet.

Ballooning has often been credited with sending pioneers into new areas of all kinds and fancied to account for spider distribution over the entire world. Such claims are now regarded as groundless. Spiders fly from local stations to nearby or even distant ones but all this flying occurs within the natural range of the species fixed by geographical and historical factors. Of notable importance have been accessible land areas over their long and changeable history, prevailing air currents and their patterns, and ecological necessities demanded of each species. In other words, spiders usually fly within their circumscribed ranges, and perish when they are dropped into areas unfriendly to their needs. Ballooning has been a principal means of colonizing oceanic islands but many groups have been obliged to arrive by rafting or trade. On the bleak cliffs of Mt. Everest, at an elevation of 22,000 feet, Hingston found tiny jumping spiders hopping about on the surface and hiding underneath stones. These could easily have been carried upward by air currents. On the other hand, it is probable that they were permanent residents living at an altitude too high for almost any other creature, and undoubtedly existing on small insects unnoticed by Hingston.

In the temperate zone aeronautic spiders are most numerous during Indian summer, when balmy days follow cool nights. In 1918, J. H. Emerton studied the aerial fauna in Massachusetts and listed sixty-nine species that took to the air during the days of his observation. A considerable number of these spiders were fully mature, others were advanced in their age, but all were of rather small species. It is now well known that many adult and half-grown spiders fly, and that this activity is not confined to spiderlings just emerging from their egg sacs. Emerton characterized the males of *Zygoballus terrestris*, a stocky little jumping spider, as being "a regular autumn flyer." Males of some of the smaller orb weavers, such as *Araneus pegnia* and *Araniella displicata*, may be seen ballooning on sunny afternoons, floating a few feet above the ground on long filaments.

It is probable that many groups of true spiders use this dispersion device during at least part of their life. Those that shun the light during all their life may not resort to flying; and only a few of the mygalomorph spiders are credited with this activity. The tarantulas are not known to balloon at all, and the large size of their young would seemingly preclude such activity. The purse-web spiders, notably the European *Atypus piceus*, are said to take to the air for short distances, so it is possible that some of the smaller four-lunged spiders also have this singular habit. A few years ago Dr. W. J. Baerg described the dispersion activities of one of the Arkansas trap-door spiders, *Ummidia carabivora*. The young leave the parental burrow and walk in single file toward and up a tree of considerable size, leaving behind them as a record of their march a silken band that can be traced back to the trap door. From the tree the plump little spiderlings "spin out a thread of silk, which, when having sufficient buoyancy, carries the spiders off and out into the world." Dr. Baerg did not see the babies fly, and perhaps they did not; in any case we know nothing of the distance they covered or of their flying behavior. Similar dispersion bands are laid down by species of *Aptostichus* in the Coast Ranges of California. However, this activity may be limited to certain species or only indulged in occasionally. The young of some Mexican species of *Ummidia* remain in the burrow with the mother until they are much too large to balloon.

The drifting threads of spider silk are known in prose and poetry as gossamer, a name of uncertain derivation but possibly from "goose summer" in "reference to the fanciful resemblance of the fragile skeins of silk to the down of geese, which the thrifty housewife causes to fly when she renovates her feather beds and pillows." The gossamer season is known in France as *fils de la Vierge*, and in Germany as *Marienfaden* or "Our Lady's threads." The reference here regards gossamer as being "the remnant of Our Lady's winding sheet which fell away in these lightest fragments as she was assumed into heaven."

Great showers of gossamer have fallen in many places in the world, and their origin has been subject to fantastic interpretations. The true explanation is a very simple one. During the autumn months, spiders become greatly active and cover the meadows and shrubbery with innumerable filaments, which soon form a thin webbing over everything. Many of these threads are put out by spiders in unsuccessful attempts to fly, and remain hanging on the vegetation. The matted gossamer is then picked up by the wind and showered down in spots often far from where the cobweb originated.

In the Yosemite Valley of California is located a series of arches which form natural traps for spider threads carried upward by the air currents and deposited in vast sheets. In these areas "all the shrubs, bushes and trees are webbed about in such a manner that the trunks of the largest trees are but faint shadows, while limbs and foliage resemble a glistening mass of crystal. In the midst of this mass are bunches of rolled-up leaves that are as white as cotton and quite thick. When the mass is disturbed by a gentle breeze, it moves throughout its entire length with a graceful undulating motion." The gossamer that falls during rainstorms in California undoubtedly has its origin in such concentrated areas of silk.

It is generally believed that ballooning and its resultant dispersion is an instinctive impulse based on necessity, and that it constitutes a protective device. The scattering of the many babies from the site of the egg sac apparently works against overcrowding and fratricide, and improves the chance of survival for each aeronaut. However, we must remember that flying is not the province solely of the spiderling, and that spiders of all ages indulge in it, limited only by size and weight. During their babyhood spiderlings eat very little and probably, in many cases, represent no great menace to each other. On the negative side, it seems certain that a high percentage of aeronauts may drown or be dropped in situations where they have little chance of survival.

THE EGGS

The life of the spider begins at the time when a zygote is formed by the uniting of the male spermatozoon with the ovum of the female. It is believed that this occurs soon after the eggs are laid by the female. The mother spider prepares a silken sheet on which the eggs are laid. They issue one by one from the genital opening beneath the base of the abdomen, and are bathed with a syrupy fluid in which quantities of sperm from the stores in the spermathecae have been discharged. At this time the eggs have a very soft chorion, which is easily penetrated by the sperm at any point.

Spiders have long been listed among animals that are able to reproduce parthenogenetically, that is, without having the eggs fertilized by the male gamete. Early records dealing with this subject have now been completely discredited. It has become well known that females can store the sperms of their mates for weeks or months, and that they are thus able to fertilize several masses of eggs in succession at distant time intervals from the product of the initial fertilization. Recent workers, however, have shown that parthenogenesis does occur in a number of instances and these are being added to each year. Some of the minute ochyroceratids of the genus *Theotima* are not known from males in spite of large collections of female specimens. Although still largely based on circumstantial evidence, it is probable that some *Theotima* species from Africa and the West Indies are parthenogenetic.

After laying a mass of eggs, the female covers them with a silken sheet and molds the mass into the egg sac characteristic of her species. The eggs are ordinarily spherical, or broadly oval, but their shape may be largely determined by their position in the egg mass. Many spiders cover the eggs with a viscid secretion, which hardens and agglutinates the mass into a single body. In some cases the eggs are only lightly agglutinated, held together in a mass by a few threads, and thus retain nearly a spherical form. Frequently, the weight of the mass is so great that the eggs assume the shape forced upon them by the available space, and thus are irregular in outline. The size of the eggs varies between rather wide limits, being 0.4 mm in some of the smaller true spiders, but attaining 4 mm in the large tarantulas.

The number of eggs laid by different spiders varies enormously. The largest of all spiders, *Theraphosa blondi*, is reputed to lay as many as 3,000 eggs, and the large orb weavers and pisaurids, which frequently spin more than a single egg sac, are credited with 2,652 in a single sac. At the other extreme we find many tiny spiders that habitually lay only one, two, or very few eggs at a time, and perhaps no more than a dozen during their lifetime. The average number of eggs for average spiders is in the neighborhood of 100. Those habitually producing more than a single sac usually place fewer eggs in each, so that the average is not greatly increased.

There is a good correlation between the size of spiders and the number of eggs they are physically capable of producing at any one time. We expect the large tarantulas to be large egg producers, and find this true, as is well shown by Baerg's average of 812 eggs per sac for one of the large southwestern American species. The contents of five sacs varied from 631 to 1,018. The sacs of tarantulas are large flabby bags often two or three inches in diameter. A still unopened sac of *Hapalopus pentaloris*, a brightly colored and curiously marked tarantula of moderate size from Mexico, contained 986 young and each of the babies measured about 3 mm in length. Only 288 eggs were found in a sac of *Phormictopus canceroides*, a very large West Indian tarantula. Another unopened sac of this same species was 2½ inches in diameter and contained 252 eggs in the deutovum or second embryonic stage. The eggs of the first sac measured about 4 mm in diameter and the deutova of the second were about 7 mm in length.

Larger eggs are produced by spiders of greater size. The eggs of *Phormictopus* are as large as small peas and exceed by several times the bulk of those of any true spider. The young of these spiders after the first molt are quite large, 7 mm, even before they have left the egg sac. It is small wonder that ballooning is not a characteristic of this group of spiders. Some true spiders produce a greater number of eggs during a single year, but female tarantulas live several years, and in total number of eggs produced probably far outdistance all spiders.

True spiders may produce few or many eggs and may place them in one or in several separate cocoons. A tiny cave spider, *Telema tenella*, lays one egg at a time. The blind spider of Mammoth Cave in Kentucky is said to lay from two to five eggs. Among the more generalized true spiders those of the family *Oonopidae* lay few eggs, and *Oonops pulcher* of Europe is known to place only two in a cocoon. The Peckhams state that *Peckhamia picata*, a small antlike spider, produces three eggs. They assumed that ants had few enemies (a supposition for which there seems to be good evidence) and that creatures resembling them would not have to produce so many offspring to maintain their normal population. Likewise, in many other families, small spiders produce few eggs simply because the abdomen is too small to accommodate many eggs, each of which must provide sufficient food for the growing embryo. They multiply their low production by maturing eggs for several distinct layings.

Medium-sized spiders produce moderate numbers of eggs. *Trachelas tranquillus*, a common eastern American species often found in houses, lays thirty or forty eggs. Many small wolf spiders produce 100 or even less. The common labyrinth spider, *Metepeira labyrinthea*, spins five or six cocoons and places about thirty eggs in each. *Uloborus glomosus* (formerly known as *americanus*) also places a string of cocoons in her orb web and leaves about fifty eggs in each. A species from the mountains of southern Arizona, *Uloborus arizonicus*, is a social spider and spins several sacs in each of which are about sixty eggs. And finally, the gregarious *Uloborus republicanus* of the American tropics, somewhat larger in size than the other species, spins larger cocoons, in which are as many as 163 eggs. The eggs of these three species are essentially the same size, measuring from 0.6 to 0.7 mm.

The large orb weavers produce several hundred eggs. The Peckhams state that the orange garden spider lays from 500 to 2,200 eggs in its cocoon, but McCook believed that 1,000 eggs was about the average number for the species. Kaston has reported 2,652 eggs in a single sac of *Araneus trifolium*, which is exceeded in size by many species of that genus. The cocoon of one of the large fisher spiders from Oklahoma, *Dolomedes triton*, contained 1,537 eggs in its large brown egg sac. The smaller *Pisaurina mira* had 518 in a sac of average size. Bonnet records a total of 2,292 eggs in the four sacs of the European *Dolomedes fimbriatus*, a species much smaller in size than several American members of the genus.

When multiple cocoons are spun by a single female, the number of eggs is less in the later ones. A female of *Nuctenea cornuta* (formerly placed in *Araneus*), which made 10 sacs, laid a total of 1,210 eggs, deposited in the following order: 234, 218, 182, 140, 112, 87, 81, 72, 51, and 33. In instances of this kind, some of the later eggs may be infertile, owing no doubt to the exhaustion of the semem supply stored in the receptacles, and perhaps also to its gradual loss of viability. In the later cocoons, the exhaustion of the female is apparent in her spinning ability, which becomes progressively less perfect. In order to maintain the normal population of a species, spiders produce a sufficient number of eggs to cope with all the factors of the environmental resistance, and emerge with a pair for each one in the normal population. The female *Argiope aurantia* lays 1,000 eggs, covers them with a tough cocoon, and yet has an average survival from the large number of only one pair. *Peckhamia picata* lays a few eggs, stationing them at different places in three or four cocoons, and still maintains an average population.

THE EGG SACS

The essential work of the female is completed when she has laid her eggs and enclosed them in some kind of silken sac. This act frequently represents the last effort of the mother in behalf of a new generation she may never see. But though early death is the lot for many, it may be delayed long enough for the mother to guard the sac for a limited period and even to aid in some way the emergence of her brood. In some species, the female spins more than one sac and must then dispose of the others in her web or hide them away, in order to give her time to the newest sac.

In the contents of the sac rest the hopes of the whole species for survival, so it is not surprising that considerable attention may be given to the fabrication of the cover. Many egg sacs are strongly made, beautifully designed creations, often pleasingly tinted with colored silk. Especially constructed for her eggs by the female spider, the egg sac is fundamentally different from an insect cocoon, which is the covering the larval insect spins around itself and in which it transforms. The degree of perfection of the spider sac is related, but only in a very general way, with the danger of destruction to which it may be subjected. When the mother spider remains with her eggs until the young hatch, the need for a tough sac may not be so great. Similarly, a sac hidden away in the depths of a burrow or surrounded by barriers of dry or viscid strands is usually not strongly made. The place in which the sac is stationed and the length of time it must remain in this location before the young desert it are the important considerations. Probably in response to such stimuli, spiders have developed different means of achieving a normal hatching of progeny under varied circumstances.

Most spiders are provided with a set of silk glands especially used for the building of egg sacs. Known as cylindrical glands because of their form, they feed their products through spigots on the outside of the posterior spinnerets. The silk spun from these glands is frequently different in color from the dry silk, and from that produced by other glands. In addition, the silk of the egg sac is different in its physical properties, being less elastic and not as strong as the dragline silk. It is apparently never viscid. The outer, varnished layers of some sacs suggest that the outer envelope is different in origin from the silk of most of the sac, or differs at least in the manner of being carded and applied as a layer.

The egg sac is generally a spherical or lenticular object, resembling a little ball, a biscuit, or a flat disk. The manner in which these sacs are produced illustrates the fact that even in realizing such commonplace structures, the spider must give considerable time, and exercise great instinctive ingenuity. Take, for example, the small wolf spiders whose sac making can be conveniently observed. Ordinarily, *Pardosa* spins a light scaffolding of lines attached to adjacent objects, and between them lays down a flat sheet of silk. This sheet usually takes the form of a circular disk approximating in diameter the length of the female. It is a closely woven fabric made by brushing the hind spinnerets from side to side and rotating the abdomen and body. The finished base may be nearly flat, but frequently it is a shallow basin, a veritable cradle for the eggs.

The actual deposition of the yellowish eggs requires only a minute or two. The gravid female stands over the sheet and extrudes through the oviduct a viscid fluid that forms a pool on the silk into which the eggs, singly or in small groups, are laid. The viscosity of the fluid is such that the egg mass largely retains its globular shape. In this fluid are sperms from the seminal receptacles. The female next spins, over the mass, a somewhat smaller covering similar in texture to that of the base, and then cuts the biscuit-shaped object loose from the floor of the scaffolding. This she now seizes and holds beneath her cephalothorax and then slowly revolves by means of her palpi and legs. At first the spinnerets sew up the edges between the two circular sheets until the break is scarcely apparent. Then the mass is revolved in all directions and the spinnerets put down additional layers of silk until, as the sac is molded and shaped, a nearly spherical object results. Soon after completion of the sac, its white silk takes on a tinge varying from gray to yellow, blue, or green; and then the spider attaches the bag to her spinnerets.

Many spiders spin this type of sac. The great flabby egg purses of the tarantulas are prepared in the burrow and are guarded by the mother until long after the young emerge. The delicate silken bags of the trap-door spiders often hang from the side of their burrow. The large lens-shaped bag of the huntsman spider is held beneath the body of the female, who will not relinquish it without a struggle. Many vagrant gnaphosids guard their eggs, but others place their tough little sacs—colored in shiny yellow, pink, or red—close against a rock or a chip of wood and then leave them.

The simplest type of eggs sacs are those of the long-legged pholcids and other primitive spiders, which use only a few threads of silk to hold the mass together. The cosmopolitan *Pholcus phalangioides* glues her few eggs (Plate X) lightly and carries the mass in her chelicerae. The tiny funnel-web tarantulas of the genus *Microhexura* also carry their eggs in this manner, and thus minimize the need for a strong sac.

Some of the most marvelous and elaborate egg sacs are spun by the sedentary spiders, which put their web-spinning superiority to good use in constructing cozy egg cradles. The sac may hang in plain view among the threads as the central theme of the web, or it may be tied nearby to herbs or other objects. Along with the special attention accorded the precious egg mass goes a somewhat different method of manufacturing the finished product.

The large, pear-shaped sac of the orange *Argiope aurantia* is hung near her web and is constructed in a most unusual manner. *Argiope* always hangs downward from the threads of her slightly inclined web, and her spinning activities are profoundly influenced by this posture. A series of cross lines attached at several points prepares a firm scaffold for the sac, which itself is a compound structure. First, yellowish threads are laid down to form a roughly rectangular roof, and on this the female spins a thick tuft of fluffy yellowish silk, which forms an irregular mass above her. Into this yielding feather bed she next spins a firmer sheet of dark brown silk, comparable to the base in which *Pardosa* places her eggs, and which serves the same purpose for *Argiope*. She lays the eggs upward against the brownish sheet by forcing the viscid liquid and the many hundreds of eggs through the genital orifice. (Most of the sedentary spiders that hang downward from webs, and even some of the vagrants that run upright, defy gravity by depositing their eggs in this strange manner.) The egg mass hangs as a yellow spherical ball, and over it *Argiope* spins a thin but tough covering of whitish or yellowish silk, which is joined to the brown silk disk. Around the whole mass—the eggs, their covering, and the rectangular roof—she then spins a fluffy covering of rusty brown or yellowish brown silk, very loosely packed, which forms a voluminous blanket around the egg mass. These lines are spun with the aid of the spider's hind legs, which comb them out of the hind spinnerets in loose loops and pat them down into the mass. Over the spongy padding *Argiope* now puts down a more finely spun covering of white or yellow silk, largely made by using the hind spinnerets alone. Smooth and closely spun, this outer covering hardens, becoming a dry yellowish or brownish cover that crackles like parchment.

The orange *Argiope* thus produces, after several hours of tireless spinning, six different sheets, tufts, or covers and from them makes three envelopes for her eggs—a thin white inner fabric, a thick woolly or flossy blanket, and a tough outer cover. The innermost layer is essentially the same as that spun by *Pardosa* and many other spiders, and is composed of two parts, the sheet that receives the egg mass and the cover.

In some orb weavers, the sac is drawn out into a short or long neck or stalk. *Mastophora* hangs her sac (Plate 17; Plate XVIII), a globular bag with a thick stalk over twice its length, on twigs and leaves near her nest. It is doubtful that the stalk contributes in any way to the security of the eggs, inasmuch as it is easily available to any insects that can reach the twigs. In some other spiders, the ball of eggs is suspended in midair by a thread of silk. The pale brown bag of *Ero*, with its irregular covering of brownish silk, hangs on its inch long pedicel in a cavity beneath a stone or under boards. The golden brown balls of *Theridiosoma* frequently are found hanging from vegetation, suspended by a long thread. It is very likely such a pendant sac offers difficulties to predators that might destroy it if it were nearer at hand.

The use of silk coverings to give the eggs a relative security from predators must have been discovered early in the history of spiders. Even a superficial silken covering would be a deterrent since many insects cannot penetrate it and might even become entangled in the threads. Spiders have, in the course of time, added many refinements to their sacs and thus gained greater protection from predators. The covering has been toughened, thickened, variegated with tufted and woolly silks and, in many cases, several blankets envelop the egg mass. Often the sac is plastered with layers of mud, or embellished with bits of wood, leaves, stones, and other debris, rendering it less conspicuous. Some sacs are glued to stones, tied to twigs, enclosed in folded leaves, or suspended at the end of fine threads. Others are stationed in the center of the capturing web or lie behind a tangle of threads in a retreat. In spite of these strategies, predators still manage to penetrate the sac and despoil or eat the contents.

Some spiders have divided the risk by putting their eggs in several baskets. They spin a series of sacs, which hang as a string in the center of their snare, or are left singly here and there. It is uncommon to find every sac in a string parasitized, whereas the whole effort of a mother spider may be lost in a single bag.

In addition to this type of protection, the spider mother sometimes plays an active role in seeing her eggs through to hatching and babyhood. The crab spiders and many hunting spiders guard the egg sac and strenuously resist, perhaps not too often with success, efforts to pilfer the contents. Wolf spiders drag their sacs attached to their spinnerets and later carry the young around on their backs until the spiderlings are able to fend for themselves. The varied efforts made by mother spiders to provide for the welfare of their eggs and young are remarkable and complex, and especially noteworthy because they are largely instinctive activities.

HATCHING AND EARLY DEVELOPMENT

Spiders undergo a development within the egg that is comparable to that of other arachnids and also of insects. The embryonic spider gradually takes form on the outside of the vast sphere of yolk that makes up most of the egg. On the generalized part, which will become the cephalothorax, appear little buds, which gradually become differentiated into the chelicerae, palpi, and the legs. A similar series appears on the abdominal portion, associated with a rather definite segmentation of eight to twelve segments, but all those behind the sixth true segment disappear as development proceeds. The basal pairs of buds persist for some time, and those of the fourth and fifth segments develop into the paired spinnerets. The buds on the second and third segments become invaginated and go to form the book lungs. Finally, the embryo nearly encircles the outside of the egg and the ventral surface is outside, unnaturally bent and convex, so it can lie within the stiff chorion. At this point occurs what is called "reversion," a process by which the position is reversed and the cephalothoracic portion becomes free. At about this time too, with the pressure against the chorion of the expanding embryo, and with the aid of a sharp egg tooth at the base of the pedipalpi, the egg covering is broken.

With the shedding of the chorion, or egg covering, there is revealed a creature somewhat spiderlike and yet obviously different from the well-known spiderling. (The term "larva" has been applied to this stage, but since that term more commonly describes insects at an active feeding stage and has quite a different sense, it will not be used here.) It is not unreasonable to suppose that this imperfect creature is prematurely hatched, and that it actually represents part of the embryonic stage. In order to gain space for fuller development and more freedom, the tough chorion is broken but the creature is still swathed in embryonal membranes. In mites, an analogous stage is called the deutovum, and is so similar to what exists in spiders that the term may be applied to the latter also.

This period in the spider's growth is not nearly so simple as was once supposed. Dr. Ake Holm has discovered and described in various Swedish spiders two, or even more, incomplete stadia (the intervals between molts), each marked by the shedding of a membrane. Some spiders hatch from the egg at a more advanced stage than do others, the degree of development being roughly approximated by the specialization of the family. In *Segestria*, the first postembryonal stadium brings to light a very primitive creature, whereas in a more highly developed spider, such as *Pardosa*, the deutovum is far more advanced.

The deutovum is without dark coloration of any kind, the carapace usually being milky white and the abdomen somewhat duller. Tarsal claws are completely lacking on the pudgy legs. The creature is unable to feed or spin, for only parts of these important structures are developed. No setae or hairs are present on any part of the body. The shape and size of the eyes are sometimes indicated even at this stage, but they are colorless and without function. In the abdomen is an abundant yolk material on which the creature can subsist until able to feed. The deutovum grows quickly after emerging from the egg covering, and is soon twice as large as the space formerly occupied by the egg. The duration of the deutovum stage is usually quite short, and toward the end of it we begin to see the darker coloration of the growing spiderling beneath the cuticle.

The first true molt, always undergone while in the egg sac, brings to light the creature that we all recognize as the spider, and which is truly a miniature of the adult. During a rather indefinite period of its life, perhaps for several stadia, it is referred to as a spiderling because of its small size. The legs are now longer, much more slender, and clothed with darker spines and hairs. At the tip of the tarsi are found two or three tarsal claws depending on the family to which the spiderling belongs. The spiderling is now able to spin but it uses little silk until after it leaves the egg sac. The digestive system is more perfectly developed, and the spiderling is probably able to feed, but its food requirements are still being met by unused yolk material in the abdomen.

What happens next is largely dependent upon the temperature. If the weather is favorable, the spiderlings become active and move about in the sac, their actions dependent upon the degree of warmth that penetrates through the silken covering of their domicile. Some female spiders guard the egg sac until they die, and others are reputed to aid their babies to escape from the sac by tearing it open. In most cases, however, the female has already died and the escape must be made by the spiderlings themselves. In tough sacs they usually cut a neat round hole, through which they emerge one by one or; in weaker sacs, they force a large rent. Following emergence comes the gradual dispersion of the family, by moving bit by bit away from the emptied sac or clustering and ballooning away.

If the weather is cold, the spiderlings in the sac are inactive. They often stay in the egg sac through the whole winter, awaiting the proper temperature of the spring before emerging and dispersing. This is particularly true of those species that lay their eggs in the late fall when not enough time and warmth are available to allow the spiderlings to develop and disperse.

MOLTING

At rather definite intervals in its development the spider casts off the bonds of its stiff outer covering and readjusts itself for a life in a more advanced stadium. method of providing for increase of size when the old cuticle becomes too tight.

This molting, or ecdysis, is characteristic of all arthropods and is ordinarily their They emerge from their transformation with shiny new armor, fully set with new hairs and spines, and often even with new structures not represented in their previous condition. In the spiders, metamorphosis brings with it a rather gradual change from the spiderling to the adult, and is comparable for the most part to the incomplete metamorphosis undergone by grasshoppers and many other insects. After each molt the epidermis formed under the old cuticle is capable of considerable increase in size before it becomes hardened. A much greater change occurs at the last molt, since it brings to light the fully developed, sexually mature adult. Only some of the more primitive spiders molt after sexual maturity; they are long-lived females that continue to live long after their first mates are dead. Apparently molts after sexual maturity are not necessary for growth, inasmuch as these females have reached their maximum size and may even decrease in size thereafter. Perhaps these molts are required to renew dull or broken chelicerae, damaged appendages, and provide a new and complete covering of hairs, spines, and other organs on the cuticle, which are the prime sensory equipment of spiders, and without full use of which they would remain at a distinct disadvantage.

Molting is ordinarily preceded by various symptoms that indicate the approach of the ordeal. This is particularly true of the later molts, which are of longer duration and more difficult of successful completion. For hours, days, or even weeks, the spider refuses to feed and becomes more and more lethargic. Certain changes in color have been noted, in some instances a darkening of the legs, in others a lightening or darkening of the whole body, owing no doubt to the changes going on under the old integument. Burrowing spiders often spin up the entrance of their burrows or block the opening with a plug of earth. Those that normally live in silken nests or leaf retreats use them for molting quarters. Some of the orb weavers hang exposed in their webs and are thus in an especially vulnerable position.

The details of molting vary little among groups of spiders, but they are of considerable interest. The large American tarantulas (Plate 5) are fine performers and their molting activities have been described a number of times. During the late summer they usually show evidences of an impending change and refuse to accept food for days or even weeks. The dorsum of the abdomen has by this time usually been rubbed completely bare of the urticating hairs, which results from the normal scraping actions of these creatures; and because they have worn this hair covering for a full year, their bodies are dull and quite bleached as compared with their fresh condition.

Tarantulas ordinarily have their quarters liberally covered with silk, but on this occasion they supplement it by spinning an expansive, closely woven sheet, appropriately termed the molting bed; this requires several hours of intensive spinning. On this soft cover the spider lies, turned completely over on its back and with legs outstretched, the front and hind ones with tarsi affixed to the silken bed. To all appearances it is dead, but if one watches the supine figure closely, occasional slight movements can be detected. After two or three hours, the old skin splits along the sides of the carapace, and the old shield comes loose from the old integument below. Splitting continues over the pedicel and the sides of the abdomen until the dorsum of the whole spider is partially freed, the old skin adhering more or less closely for some time. At this stage, the spider has changed its position so that it is now lying on one side, and it then begins the laborious process of pulling the appendages from their old casings. The spider extracts its chelicerae first, and then begins a series of rhythmical contractions that gradually bring to light the femora, patellae, and successively the rest of the legs. The front legs and the palpi are freed initially, then come the posterior legs. After about an hour, the cephalothorax and the legs are completely freed; whereupon the spider easily extracts the abdomen and moves away from the cast skin. For three or four hours it lies on its back or side while the new skin hardens; then it resumes its normal upright position. The freshly cast skin of the tarantula is moist inside, and the new cuticle also shows traces of moisture. The molting fluid between the old and new skins aids the progress of the molt by loosening the old skin. The cast skin includes all the ectodermal cuticle and in the case of mature females the old seminal receptacles which are now replaced by new ones.

Essentially the same picture is presented in the molting of true spiders. The sedentary spiders hang in their webs or in their retreats. Many vagrants spin a few threads in a favorable nook and hang downward, their tarsi fixed in the silken lines, and their abdomen supported by a thread from the spinnerets. The cuticle splits around the sides of the carapace and around the abdomen. Then the legs are freed in slow stages by the usual rhythmical contractions, the front legs coming out first, and finally the posterior pair. By the time this is accomplished, the abdomen is virtually free, and the whole spider is suspended in the air by the thread from the spinnerets. The whole process requires ten or fifteen minutes, the length of time apparently being governed by the size of the spider. Young spiders suspend themselves, molt, and lengthen their legs in less than half an hour, the molting itself often taking only three or four minutes. Half-grown spiderlings require about an hour, and young males and females in the last molt require about two hours. Molting may occur during either day or night, and seems not to be limited by time as are many of the other activities of spiders.

The freshly molted spider is much paler and softer than in the previous instar, and only gradually hardens and darkens its new integument. During this relatively brief period occurs the entire increase in size of the carapace and appendages until the next molt. Size increase is usually progressive and is determined by the instar and the sex of the spider. In some species the legs increase tremendously in length between the instars. Growth commences as the legs are pulled out of the old integument—much as fingers are pulled from a glove. It continues during the time the spider is free of the cast skin and hangs suspended. The appendages are bent back and forth in a regular ritual. Pierre Bonnet has demonstrated with some very ingenious experiments the necessity of these calisthenics following the molting. Without such bending movements the appendages become sclerotized even at the joints and remain stiff.

The number of molts necessary to attain maturity varies widely in spiders. Bonnet has shown rather conclusively that size is the deciding factor in most species. Tiny species molt few times, whereas large ones molt a greater number of times. Bonnet noted that small species of 5 or 6 mm in length (*Pholcus phalangioides*, *Uloborus plumipes*, etc.) molted four or five times. Species of medium size, measuring about 8 to 11 mm (*Araneus diadematus*, *Pirata piraticus*, etc.), molted seven or eight times. The larger spiders, 15 to 30 mm in length, molt ten to thirteen times (*Dolomedes plantarius*, *Nephila madagascariensis*, etc.). The largest of all spiders, the tarantulas, molt even more often: according to Dr. W. J. Baerg, twenty-two times for the male of an American *Aphonopelma* (called by him *Eurypelma californica*). Further, in these spiders many molts after sexual maturity make it probable that the females molt between thirty or forty times before they die. At the lower limits, only four molts are credited by Bonnet to the pygmy male of *Nephila madagascariensis*; and I have discovered that the pinhead-sized male of *Mastophora cornigera* may undergo only two molts before becoming mature.

Even within the same species there is variation in the number of molts. Bonnet found that to become mature, females of *Dolomedes plantarius* molted as few as nine or as many as thirteen times. The number of molts was to some extent correlated with size, the larger individuals requiring more molts, but various other factors were important, the amount of nourishment being one of these factors. The males of this same species became adult after nine, ten, or eleven molts, a number similar in the lower limits to that of the female that they resembled in size. In the northern United States the egg sacs of the species of *Mastophora* are broken open early in the spring and the young disperse. The males emerge either in the penultimate stadium or fully mature, in the latter case having molted only twice. The females are of the same size, and presumably have also undergone one or two molts, but they must molt seven or eight times before they are sexually adult. The difference between molts is probably five or six, and reflects the enormous disparity between the size of the sexes, a difference greater than that in any other spiders known to the author. Time is also important, and whenever maturity is reached quickly for the species, the molts are near the minimum for the species. When maturity is retarded for some reason, more molts are undergone. Abundant food diminished the number of molts, whereas starving increased the number of molts.

For very few North American spiders is the number of molts known. The black widow, *Latrodectus mactans*, has been studied rather carefully by several investigators, and we find the usual considerable differences between the sexes as regards molting behavior. The males become adult after the fifth, sixth, or seventh molt, whereas females are adult after the seventh, eighth, or ninth molt. In the related spider, *Steatoda grossa*, the males become adult at the sixth or seventh molt, the females at the seventh or eighth molt. This number of molts is about average for spiders of this general size, and the males almost invariably molt at least one less time than the females. In 1927 Gabritschevsky recorded the time intervals between molts for *Misumena vatia* as part of his paper on the change in pigmentation of that species. The synopsis is as follows: deposition of eggs, July 28; hatching, August 8 (my estimate); first molt, about August 12; emergence from the egg sac, August 14; second molt, August 24; third molt, September 23; fourth molt, September 23; fifth molt, October 17; sixth molt, January 5; final molt, a time after January 5 that was not indicated.

Various morphological changes accompany molting; some of them are very significant. The presence of a third claw on the tarsi of very young spiders that are two-clawed as adults indicates that the three-clawed condition is the primitive one. Young wolf spiders have the eye formula of the *Pisauridae*, a fact that corroborates our belief that the former were derived from an ancestor very like recent pisaurids. The young of *Tibellus oblongus*, an elongated crab spider, have the general body form and the eye relations of species of the more conservative *Thanatus*.

Each molt represents a crisis in the life of the spider, and brings with it dangers of many kinds. During the transformation the spider is completely helpless, trussed up in old worn clothing, and exposed to attack from many enemies. Crickets, sowbugs, mealworms, and other omnivorous animals, not serious adversaries under normal conditions, are liable to nibble and kill the defenseless creature; it lies vulnerable to attack from the meanest foe. Other normal enemies find the spider completely unable to fight back. Furthermore, the mechanical difficulty of extracting its appendages may prove insurmountable, and the imprisoned creature will parish, or so mutilate its legs that its chance for life in a hostile world is much diminished. G. and E. Deevey found that nearly half the deaths before maturity (thirteen out of thirty-one) among the black widow spiders they reared were the result of failure to complete a molt. One out of every twelve spiders in the total of 158 that were studied from hatching to death died from this cause.

AUTOTOMY, AUTOPHAGY, AND REGENERATION

The spider shares with many other arthropods the ability to drop an appendage without great inconvenience—called "autotomy"—and the ability to replace it in more or less perfect form by subsequent regeneration. This latter power of replacing lost or mutilated organs is a very old one and, most strongly developed in lower animals, serves as a device of great importance from the viewpoint of protection and survival. Often the spider is able to escape the clutches of an enemy without greater loss than the shedding of one or two of its appendages. Whereas autotomy occurs in spiders of all ages, the regeneration of new appendages is limited to young spiders that have not stopped molting, or to those few primitive spiders which molt after sexual maturity.

It is now known that autotomy in the strictest sense—that is, the act of reflex self-mutilation—does not occur in the *Arachnida*. An appendage is dropped only after a visible effort on the part of the spider, which struggles with such violence that the tension on the member snaps it off at its weakest point. This action was termed "autospasy" by Pieron in 1907; it involves the breaking of the appendage at a predetermined locus of weakness when pulled by an outside force. This locus is between the coxa and the trochanter in the legs of most spiders: a point found by Wood to resist only 7 percent of the stress that the next weakest juncture, that between the metatarsus and the tarsus, could withstand. In harvestmen, the weakest point is between the trochanter and the femur, and in other animals the break may occur in quite different locations.

The reaction of the spider to the loss of appendages varies considerably. The loss of one, two, or even three legs in some active crab spiders seems to result in little inconvenience to the animal, which runs away without crippling effects. Stocky crab spiders that have lost the first two pairs of legs take up a position in which the short third legs are directed forward as in normal posture, and are able to move about with relative ease. The stout front legs of the crab spiders are at the same time organs of touch and offensive weapons, and when they are lost, the ability to capture prey is seriously impaired. Mature males that have lost some of their long front legs are at a distinct disadvantage during courtship, and fall easy prey to females not willing to meet them.

Autotomy is easy to observe. If a spider is grasped by one of the legs and the animal has a good hold on the substratum, the leg will break loose at the usual locus between coxa and trochanter. On the other hand, if the spider is held in the air and is unable to exert some countering force by grasping an object, it is unable to drop a leg. When held in a pair of forceps, the animal usually twists around, grasps the forceps, and literally pulls its body loose from the leg. The speed with which this is accomplished varies with the species and with the thickness of the appendages, but it is practically instantaneous once the spider begins to effect an escape. If two legs are held firmly, some spiders break both of them easily, but some of the stocky crab spiders are unable to exert enough force to free themselves.

Sometimes the spider is seized by a predator that is able only to break the cuticle of the leg, with the result that blood begins to flow through the break. Although the spider may escape otherwise unscathed, this is a most serious situation, inasmuch as an open veinal system allows the draining of blood from the body. The instinct of the spider is immediately directed to a preventive device: the leg is pulled out, and the flow of blood is quickly halted at the normal breaking point between the coxa and trochanter. This amputation is accomplished with the help of the remaining legs and the mouth parts. In some instances the spider spins threads and ties the appendage to them, and is able to amputate a whole leg or even a small stump and thus save itself from almost certain death.

Autotomy can be put to use by the spider to rid itself of an appendage that is unwelcome for some reason other than injury. The known males of the species of *Tidarren* have long been observed to carry only one palpus, a great bulbous affair held in front of the head. In the antepenultimate stadium, the palpi are only slightly swollen but after molting the male has two tumorous enlargements resembling boxing gloves. So large are these new members that the spider is handicapped by them, and able to manipulate them only clumsily. The obvious solution to the problem is the amputation of one of these palpi, and this is exactly what the spider does by a most interesting process. It spins a scaffold of silk, similar to the molting sheet, and, suspended from it by its legs, fixes one of its palpi in the threads. The spider now twists around and around and, aided by pressure from its hind legs, twists off the unwelcome palpus. The spider now has a single palpus which is held in front of the head and occupies much of the available space. At the next molt the male becomes sexually mature, and the vital parts of the palpus are revealed. A second palpus is never regenerated to replace the lost one.

The spider's instinct to rid itself of an injured or inconveniencing appendage takes precedence over all other instincts, but once autotomy is accomplished the spider almost invariably does a most curious thing. It picks up the bleeding member and sucks the juices from it, usually discarding it only after it is sucked dry. This "autophagy" is perhaps as old a habit as autotomy itself, but this act may not have any especial significance beyond its general interest. Spiders often attack each other, or other prey, and if successful only in capturing a leg, will stop and suck on it in the same manner. The instinct associated with bleeding prey and the taste of blood prompts the feeding action.

If a leg is lost by an immature spider, it is replaced by a smaller, imperfect replica at the next molt, provided a sufficient time has elapsed between the loss and the molt. The regenerated appendage increases in size with successive molts but never quite attains the full perfection of the normal appendage. The same leg can be regenerated repeatedly, so long as the spider is still immature, but at least three successive molts are necessary to attain a size comparable to that of the normal appendage.

The regeneration of a leg takes a definite course. A good illustration is that of the crab spider, *Misumena vatia*. The females of this species have white legs, and when one is lost, the appendage that replaces it is shorter, unmarked (as is to be expected), and deficient in the number of spines, as compared with the normal appendage. In the male, however, the first two pairs of legs are banded after the third or fourth molt, and in each successive molt the amount of pigment in the dark annulae increases. If the male loses a leg during the third instar, when it is still white, after the next molt the regenerated leg is wholly white, but the normal front legs continue to increase their annular pigmental areas. After the fourth molt, the regenerated leg becomes annulate, but the depth of the color is much less than that of the normal leg. In other words, in *Misumena vatia*, a regenerated leg takes on the normal coloration of the leg at a previous instar and never quite approximates the normal leg in size and color.

LONGEVITY

Most spiders that inhabit the temperate zones live only one year. The yearly population may be divided very roughly into two faunas, one identified with the spring and the second with the fall. Over-wintered or recently matured males of many vagrant spiders are abundant in early spring and are on hand when their females become mature. The crab spiders are found on the ground, in the corollas of spring flowers, or running over the stems of shrubs and the bark of trees. The grassland teems with jumping spiders, wolf spiders, clubionids, and many other wandering types. The sedentary web spinners are likewise well represented by many species that spin inconspicuous orb webs or tangled webs on the vegetation or in ground detritus. In a few weeks most of the males disappear and gravid females are found on all sides, some spinning up domiciles for egg laying while others, having already made their first sac, carry it attached to their spinnerets or held in their jaws. By midsummer the possibility of finding males of the spring spiders is not very good, but the females carry on far into the year, often laying several egg masses. Thus, in the spring and early summer we have the males and females of the spring fauna, and the juvenile and growing representatives of the fall fauna.

In the fall, the sedentary spiders advertise their presence in a conspicuous manner by great sheet webs and expansive orbs. The males appear in midsummer and early fall and attend the females on the outskirts of their modest webs. In August we find the webs of the grass spiders on grass and shrubs, grown much larger by accretion, and can surprise the adults pairing in the funnels. The orb weavers now are attaining maturity in great numbers, and every suitable situation is filled by a web of variable dimensions. Having vastly increased in size, they spin much larger webs. As the season progresses, the males dwindle rapidly; soon all are gone after having lived the shorter, intenser life identified with their sex. The females lay their eggs and enclose them in sacs of various kinds, which tend to be more substantially built and more heavily insulated than those of the spring spiders. After the killing frosts in November, most of the adults of the fall fauna are gone, but already the growing young of the spring spiders have attained nearly their full development. During October and November, the partially developed spring spiders engage in ballooning activities in company with many precocious fall spiderlings. The fall spiders produce their eggs in the fall, and their young either spend the cold winter months in their sacs or, having deserted them, they live under debris or in protected places until warmer days allow them to begin their march to full maturity.

It must not be thought that the two faunas are discretely separated one from the other. Actually they are bridged by species that mature during the midsummer. Because of multiple cocooning and precocity, or tardiness of some species, there is a considerable overlapping of the faunas. Further, some species do not seem to conform to any definite pattern, and mature males and females may be found during almost any month of the year. In the American South it is possible to have two full generations during the year. For the most part, however, even in warmer areas we find only one generation each year. Over much of Canada there is only one generation per year, while in the colder northern reaches and in the high mountains some of the species require two or more years to attain complete maturity.

The life of the male is invariably shorter than that of the female. Males of the spring spiders die in early summer after having lived about ten months from the egg to adult. The same longevity holds true for the fall species, and the females outlive the males by weeks or months. Under laboratory conditions G. and E. Deevey found that average male black widows matured in about 70 days and lived a total of about 100 days; whereas females matured in 90 days and lived about 271 days. The greatest life span for the males was 160 days, and for the females 550 days. Such data show that mating between brothers and sisters of the egg masses is quite improbable.

A number of spiders are known to live more than one year. In the northern United States almost the only ones to do so are some large wolf spiders, which burrow into the soil and probably live two or more years. Occasionally other spiders live about eighteen months, such as the large fisher spiders and the black widows, even though their normal life span may be only a year. Laboratory specimens often persist far beyond what is presumed to be normal life span.

The more primitive true spiders often live more than a single year. Some of the segestriids and scytodids are said to be perennials, and Dr. Lucien Berland kept a female filistatid for ten years. It is probable that all ancestral spiders were longer lived, and that one of the sacrifices of the modern true spider for the many advantages it enjoys is a drastic reduction in life span.

It is generally believed that all mygalomorph spiders—the tarantulas and their many allies—live several years. The purse-web spiders are reported to live as much as seven years, and the true trap-door spiders are also perennials. Exceeding all other spiders in length of life are the large tarantulas. Dr. W. Baerg kept a female tarantula for more than twenty years and believes that twenty-five or even thirty years probably represents the normal age of some females. He found that the males of some of our southwestern tarantulas matured in eight or nine years, but they ordinarily die a few weeks or months afterward. Much more precocious are some of the giant tarantulas of Brazil, which attain full maturity in three or four years. Much remains to be learned about the time spans for maturity and longevity of our mygalomorph spiders.

THE ENEMIES OF SPIDERS

The innate biotic potential of the spider allows it to reproduce and maintain a normal population in the habitats available to it over its natural range. Antagonistic to this are all the physical and biological factors that the spider must endure to survive. Much environmental resistance is centered in climate in a broad sense which, often in changeable or even catastrophic fashion, moves to decimate a population. Destruction by flood, earthquake, fire, and drought are misfortunes common to all plant and animal life. The effect of drought was noted by W. J. Baerg, who claimed that the drop in normal rainfall from nine to three inches caused the extermination of a large colony of Mexican tarantulas. A temperature-humidity imbalance is a powerful lethal agent for many animals. The first hard frosts in the autumn usually kill off all annual spiders, notably the large araneids, but untouched are all the developing and mature perennial species presumably genetically prepared for these occurrences. Whereas the sum of all physical and biotic factors are potential enemies, here we can make only brief mention of the principal biotic opponents of spiders. In other places in this book mention of some natural enemies already has been made.

One of the most insidious and devastating of spider enemies are the fungi that infest the bodies of spiders and kill them. Victims in various stages of infestation are often seen in the field. W. S. Bristowe has described the spider devastation inflicted by such fungi in England where cold and humidity work hand in hand with these averse plants. Other plant enemies also exist, such as the insect-eating plants that claim their quota of spiders. Of similar effect are various nematode worms that grow in the bodies of spiders and, after feeding on the contents, leave the dying araneid and emerge as free-living adults. Preserved specimens often show parts of such worms protruding through the body wall of the hosts, stalled in futile efforts to escape. These internal parasites are far more common than generally believed and largely go unnoticed.

The vertebrate enemies of spiders are numerous and range through all groups from fish, amphibians, reptiles, birds, and mammals, even to the occasional tidbits claimed by some native peoples. Many spiders have been named from partially digested specimens taken from the stomachs of toads. Insectivorous and omnivorous birds eat many spiders during their foraging; notable among these are our humming birds which also make use of the spider's silk for their nests. Among the mammals that feed on spiders and dig them out of their burrows are the many omnivorous types, the shrews, skunks, coatimundis, etc. In New Jersey the *Geolycosa* wolf spiders living in burrows in my poorly kept lawns were decimated by digging skunks. On Barro Colorado Island in the Panama Canal Zone, the coatimundis dug out many of the large tarantulas of the study colonies.

The cost in lives due to the spider's vertebrate enemies is minor when compared to the loss of spider lives through its many insect enemies. Spiders are largely insect eaters and antagonists, so it is natural that the insects should turn to their prime enemies for part of their food supply; these superior arthropods are predators of spiders in all stages from egg to adult. Tiny predaceous flies and wasps invade the egg sacs of black widows and many other spiders and rear many examples of their own kind from the eggs. Notable egg sac parasites are various *Mantispa* that as young larvae invade and develop in the egg sacs of lycosids, agelenids, and other spiders, and emerge as mature adults. The small-headed flies of the family *Acroceridae* are, largely if not exclusively, parasites of spiders. The female fly broadcasts a large number of eggs onto a surface favorable for the discovery of a spider host into which the tiny larvae can bore. The larva of *Ocnaea smithi*, a large, yellowish fly with banded body, lives in the body of the California trap-door spider, *Bothriocyrtum californicum*, feeding on the tissues until as a much enlarged larva it emerges to transform to an adult fly, leaving behind the shriveled remains of the spider. Again it can be said that this parasitism is far more common than generally believed but mostly unnoticed.

Wasp predators are usually believed to be the worst enemies of spiders. The *Pompilidae* are called spider wasps because they use spiders as food to feed their young. Some spider wasps partially disable a spider by stinging it, then lay an egg on the abdomen to be carried around by the recovered spider. The larva soon hatches, gradually consumes the body of the host, and then leaves to pupate and become an adult wasp. Many other pompilids find a spider, sting it until it is parasitized, dig a burrow for it, and then deposit an egg on its body before filling in the burrow. The tarantula hawks of the genus *Pepsis* are among the largest of these wasps and for the most part prey on tarantulas. Various sphecoid wasps (family *Sphecidae*) mass provision their nests with enough paralyzed spiders to enable the young to develop to maturity, then lay an egg in each of the cells. Many fill the branches of live or dead trees, and assign an average of eight spiders to each cell of the nest. *Trypargilum tridentata*, of our southern and western states, provisions each of its several cells with an average of 23 spiders; 20,000 spider prey of this species were noted in the study of one investigator. These wasps seem to prefer web spiders and rarely take others. Some of their relatives, the thin-waisted mud-dauber wasps, make nests of mud and attach them to stones or to the ceilings or walls of buildings. Each of their cells (Plate XXIII) is filled with paralyzed spiders but these usually are a mixed bag of web spiders and vagrants.

In summary, then, it can be seen that many animals are enemies of the spider and destroy them in great numbers. Perhaps, as W. S. Bristowe believes, spiders are their own worst enemies since they often are cannibalistic and eat their own eggs, their young, or their neighbors of both sexes.

4
Silk Spinning and Handiwork

SPINNING CHARACTERISTICS

The maiden Arachne, daughter of Idmon of Colophon in Lydia, became widely known for the excellence of her work at the loom. Indeed, her art was so superb that the nymphs from the woods and streams came to gaze upon it. Many wondered whether even Athene, goddess of weaving and the handicrafts, could surpass this maiden, who seemed to have been tutored by the gods themselves. So confident became Arachne in her amazing skill that she challenged Athene to compete with her. Although affronted by the presumption of the girl, Athene accepted the challenge and wove a tapestry showing the warfare of the gods and the fate of those who conspire against them. Arachne depicted the love adventure of the gods with such exceeding perfection that the goddess, unwilling to admit that so high a degree of excellence could be attained by a mere mortal, became enraged and destroyed it with a blow from her spinning shuttle. The rash and humiliated Arachne attempted to hang herself, but the noose was loosened and became a cobweb, and the maiden was changed into a spider. Thus disgraced, lying on the rent pieces of her tapestry, Arachne was condemned to perpetual spinning.

The Greek word for spider is *arakhne* and commemorates the weaving skill and mythical fate of the imprudent maiden. From it originated the group name *Arachnida*, which embraces all the arachnids or spiderlike creatures. From the derived Latin *aranea*, spider or cobweb, come the ordinal names of *Araneae* or *Araneida*, exclusively used for spiders. Probably in deference to the girl Arachne, the Latin languages mostly use their stem word in the feminine gender, such as *araña, aranha, araignée*, but the Italian *ragno* is masculine.

The English word "spider" is a corruption of "spinder," one who spins, and is similar in form to other Teutonic words derived from the same root, such as the

Spinne of the Germans. This root persists in different form in the words "spinstress" and "spinster," both having reference to women who spin as a profession, but the latter has acquired a quite different connotation. Finally, from the Old English *attorcoppe*, spider, later shortened to *coppe*, comes our word "cobweb" meaning simply spider web.

While most people associate spiders with a silken web of some sort, few are aware of the dependence of these animals on silk. The ability to spin is an early gift to the spiderling and is developed after the first molt and before emergence from the egg sac. Immediately upon leaving the sac, the spiderling strings out its dragline threads and attaches them at intervals to the substratum. Thereafter it is never free of this securing band through its whole life, except by an accidental breaking of the cord.

The degree of reliance on silk varies considerably among spiders. The very oldest ones, the precursors of those few we know from Carboniferous rocks, probably had clumsy appendages that were only beginning to be used to comb out a liquid silk. The most primitive of recent spiders are sometimes said not to spin a formal dragline, although they are otherwise as well equipped for spinning as some of their relatives, as is evidenced by their well-made egg sacs and silken tubes closed with a trap door. The familiar jumping spiders and wolf spiders, so often seen running over the ground or climbing on plants, are vagrant types in which the use of silk is limited. They use it chiefly for their draglines, for covering their eggs, and for lining their retreats. On the other hand, a vast majority of sedentary spiders are strongly dependent on silk. Some of them have become slaves of elaborate webs and are nearly helpless when not in contact with them. For spiders of this type silk is of paramount importance during their whole life span.

The majority of spiders are inveterate spinners and far surpass all other animals in the variety and excellence of their weaving. Some of the other arachnids produce silk, but they use it in a very limited way. The pseodoscorpions have cephalic glands and spin silk through a tiny spinneret located on the tip of the movable finger of the chelicera. Before laying their eggs, these tiny animals build an ingenious little domicile made of small particles cemented together with silk, and lined inside by a covering of silk. A few of the mites also have silk glands and are said to spin threads so fine that they are invisible to the naked eye. The so-called "red spiders" are mites of the family *Tetranychidae* which cover the leaves of trees with silk and use it as a protecting blanket for their eggs and young.

Many insects spin silk and in such profusion that they rival the work of even the sedentary spiders. The unsightly webs of the tent caterpillars are familiar and despised objects to most people but, looked at objectively, they are quite wonderful fabrications. Their tent nests are not far different from some made by gregarious spiders. Many other moths spin silk, but its use is largely restricted to making the cocoon. The most noted insect spinner is the silkworm, the larva of the moth *Bombyx mori*, which has been domesticated for so long that it cannot now maintain itself in the wild state. It produces cocoons that are easily unwound, and supplies the bulk of commercial silk. The silk of moths, caddis flies, and sawflies is produced in cephalic glands, which pour their contents through a single opening in the lower lip. The threads are usually much thicker than those of spiders. The silk is probably of only one kind.

The spider's reliance on silk is well illustrated by the many different uses to which it is put. A list of some of these uses is given below, without any attempt at other than a general classification:

Protection and Retreats

The dragline; the bridge line; the trap line of the orb weavers; the warning
threads of *Ariadna*, *Liphistius*, etc.
The ballooning lines
Attachment disks to anchor the lines
The cells and retreats of all spiders
Hibernating chambers
Molting threads, beds, and chambers
Trap-door covers; spinning up of burrows and open retreats

Protection of Eggs and Spiderlings

The egg sacs
The nursery webs of the *Pisauridae*

Web Structures Associated with Mating

The sperm web of the males
The bridal veil of crab spiders and other vagrants
The courtship and mating bowers of the black widow and sedentary spiders
The mating chambers of the vagrant spiders

Structures for Stopping and Ensnaring Prey

Sheet webs
The stopping tangle webs of the grass spiders and the aerial sheet weavers
(*Linyphiidae*)
The viscid or entangling webs of the orb weavers and some other spiders
The viscid ball and pendulum line of *Mastophora*, *Dicrostichus*, and *Clad-
omelea*

The viscid hackled band in the diverse capturing webs of the cribellate spiders
The catching thread of *Miagrammopes*
The retiarius of the *Dinopidae*

Bands for Binding the Prey

The swathing band of the orb weavers
The swathing film of the comb-footed spiders
The swathing band of *Hyptiotes*
The entangling ribbon of the *Hersiliidae*
The capturing band of *Drassodes* and other vagrant species

The above requirements and others not listed are met by the production of different kinds of silk, which are used, seemingly at the will of the spider, either separately or in combination to provide the special threads, desired bands, or drops for a particular project.

THE SILK

The silk of spiders is a scleroprotein which is produced as a liquid in varied and voluminous abdominal glands. When drawn out of the spinnerets, the liquid ordinarily hardens to form the familiar silken threads. It is believed that the mechanical stretching of the silk during the drawing of the lines is responsible for the hardening, rather than exposure to air or any chemical process. Viscid silk is produced in some of the glands and remains sticky for long periods. An analysis of the silk has shown that it is a complex albuminoid protein quite similar to that produced by the silkworm, although this similarity is denied by some investigators. The silk of the silkworm comes from modified salivary glands located in the head, whereas the silk of the spider is derived from transformed coxal glands in the abdomen.

Spider silk is noted for its strength and elasticity. The tension necessary to bring a compound thread 0.01 cm in diameter to the breaking point was once found to be eighty grams. The considerable tensile strength, which is said to be second only to fused quartz fibers and far greater than steel, goes hand in hand with great elasticity. Some threads will stretch their full length or even more before they break.

The strength of the threads is to some extent dependent on the manner in which the spider draws them out, with greater speed increasing the strength of the thread. When they are drawn speedily, the fibroin chains attain a maximum orientation, which contributes greater strength to the lines. The cocoon silk of the silkworm is essentially equal in strength to that of the orb-weaving spiders.

However, spiders produce several varieties of silk, and some differences are found among them in strength and elasticity. The viscid line of the orb-weaver snare is not very strong but is extremely elastic; whereas the foundation lines of these webs are of great strength, exceeding even that of the cocoon silk.

Most spider threads are not single fibers although they may appear so to the naked eye. The dragline thread readily separates into two rods of equal thickness, sometimes even four, but often elements from other glands lie parallel to these elementary strands and mar the uniformity. Under ordinary magnification, single fibers are rather uniform rods, but when photographed by the electron microscope at 35,000 diameters even the finest threads are not completely uniform, and show tiny enlargements and irregularities. Not much detail of the internal structure of the silk can be seen even at this great magnification. The finest single fibers attain a thinness of 0.03 micron, or about one millionth of an inch, and are invisible to the naked eye. Much thicker threads are relatively large, being 0.1 micron, or one-quarter millionth of an inch in thickness. Many molecules are larger than the width of these spider threads. It is possible that the spider can draw out its filaments to a degree equal to the thickness of its protein molecule, and that the finest threads represent a single chain of molecules.

THE SILK GLANDS

The silk glands of spiders are secreting organs located within the abdomen. Differing in size, form, and location, these organs are classified largely on the basis of their physical characters. Thus, the pyriform glands are pear-shaped, the aciniform are berry-shaped, and the other glands are similarly identified by their contour. At least seven distinct kinds of glands are known to occur in the whole group of spiders, but not all of them are found in any single family. The cribellar glands are found only in spiders that have a cribellum—a flat spinning plate— and are used in conjunction with the calamistrum, a comb of hairs on the hind metatarsi. The comb-footed spiders of the family *Theridiidae* possess all six of the remaining types of glands, and are the only ones having lobed glands, which secrete the material of the swathing film. These spiders thus are provided with one more set of glands than their close relatives, the sedentary orb weavers and the linyphiid spiders.

Even the oversimplified classifications of Apstein and others demonstrate conclusively that the spinning organs and glands of spiders are the most complicated structures known for the production and utilization of silk. The several types of glands and the uses of their silk products are enumerated below:

1) The *aciniform*, or berry-shaped glands. These glands are found in all spiders and are characterized by their nearly spherical shape and resemblance to various berry fruits, such as a raspberry. Four clusters, each containing from a

few to as many as a hundred glands, send the silk through each of the posterior and median spinnerets. The swathing band is a product of these glands. According to Apstein, these glands also produce the ground lines for the viscid drops.

2) The *pyriform*, or pear-shaped glands. Also found in all spiders, these glands occur in two clusters of a few to one hundred or more, and communicate with the front spinnerets. The making of the attachment disks is one of their functions, but they sometimes contribute wild threads to the thicker draglines.

3) The *ampullate*, or bellied glands. Known in all spiders, these usually are present as four large, long, cylindrical glands, but frequently there are six, eight, or even twelve. They open through spigots which, when four glands are present, are located on the inner side of each of the front and middle spinnerets. Most of the dry silk of spiders, the dragline being the chief agent, is produced in the ampullate glands. Comstock has suggested that the ground line of elastic silk in the orb weavers is produced by these glands, two of which have been modified for the production of this important element. The fact that the yellow silk of *Nephila* is spun from the anterior spinnerets partially confirms this opinion.

4) The *cylindrical* glands. These long, cylinder-shaped glands are often wanting in males, and are lacking in the *Dysderidae* and *Salticidae*. They number six or more, and open on the inside of each posterior spinneret through a spigot. They produce the silk for the egg sac.

5) The *aggregate*, or tree-form glands. There are six of these irregularly branched, compound glands, opening on the inner surface of each posterior spinneret through spigots. From these glands, which are found only in the *Araneidae*, *Linyphiidae*, and *Theridiidae*, are produced the viscid drops for the viscid lines of the web.

6) The *lobed* glands. Found only in the *Theridiidae*, these are irregular in shape and lobed, opening on the posterior spinnerets through spigots. The swathing film of this family is produced in these glands, which are developed largely at the expense of the aciniform glands.

7) The *cribellar* glands. These numerous, spherical glands open on the cribellum through many tiny pores. They occur only in the cribellate spiders, and secrete the woof of the hackled band.

In addition to the various internal glands of the abdomen, there are present in some male spiders of a number of families small *epiandrous* glands on the area in front of the genital groove. Silk is delivered through small spigots and it contributes at least in part to the sperm web.

As is to be expected, those spiders that use many types of silk have the greatest number and volume of glands. The abdomen of the sedentary orb weavers is largely filled by glands; whereas the vagrants are less bountifully supplied. In some males the cylindrical glands are missing, and in many males the other glands are less well developed than in females. Inasmuch as the male's need for some types of silk virtually ceases when he becomes adult, the lack of specific glands is of no great importance.

The spider has at its command these various types of silk glands and can call upon them for its many needs. Flexible fingers are the spinnerets: they can be extended, withdrawn, compressed, and manipulated like human hands. The filaments produced are sometimes simple threads in multiples of two, but more often they are composite lines and are drawn from different glands. The viscid spiral of the orb-weaver snare, for example, is composed of a double ground line, possibly coming from the aciniform glands, on which is superimposed a thin coating of viscid silk from the aggregate glands. Only when this line is spun in a particular way does it take on the characteristic form of a beaded necklace. The spiral is spun rather slowly, and then the spider pulls out the coated line and lets it go with a jerk. As a result, the fluid is arranged in globules, spaced along the line and far more sticky than a thin, uniform covering. The rate of pull and the degree of tension determine the finished product. The spider spins leisurely or swiftly, according to its need.

THE DRAGLINE

No better illustration of the dependence of spiders on silk can be presented than the habit of laying down a dragline or securing thread. Wherever the spider goes, it always plays out behind from its spinnerets a silken line, which is anchored at intervals (by means of the attachment disks) to the substratum, much as the climber lets out a rope when he enters the recesses of a deep cave or rappels down the slope of a precipitous mountain. The dragline (Plates XXIX, XXX) is a constant companion of spiders of all ages and all kinds except, perhaps, members of the primitive family *Liphistiidae*. It is the fundamental thread of most spinning.

The sedentary orb weaver, committed largely to an aerial life in the confines of its web, outlines the zones of its snare with this thread. Long strands floated in the air form bridge lines from tree to tree or across streams. On draglines, the spider balloons for long distances. Great sheets and flakes of gossamer are mostly the discarded draglines from many spiders. The orb weaver again, hidden in its leafy retreat, holds a trap line and uses it to detect the presence of an insect in the web.

The dragline is the lifeline of the spider. It is an aid in preventing falls from precipitous surfaces, and may also serve as a means of escaping from enemies. Web spiders often drop from their web on these lines and hide in the vegetation. Or they drop down and hang suspended in midair until the danger is past, whereupon they climb back up hand over hand to their original position. The hunting spiders jump headlong over cliffs or leap from the sides of buildings to escape capture, and float down gently on their silken ropes. Most of the spinning in our houses is dragline silk, which soon is transformed into the familiar cobweb, heavy with air debris. Even the framework of the retreats is put up with dragline silk, and then on this base other types of silk are laid.

Not a single filament, as the name implies, the dragline in its simplest form is composed of two relatively large threads that adhere so closely together that only one is apparent. On occasion, the dragline may be made of four strands, or even of a great many threads drawn from several spinnerets.

SPIDER THREAD IN OPTICAL INSTRUMENTS

The use of spider silk for reticules in various optical instruments is a direct consequence of the fineness of the fibers and of their great strength and ability to withstand the extremes of weather. Prior to World War I, spider silk was used very extensively for cross hairs and sighting marks in a great variety of engineering, laboratory, and fire-control instruments. For transits, levels, theodolites, astronomical telescopes, and many other optical devices there is nothing much superior to spider silk. Most people who use such instruments are familiar with the fibers, and often replace them in the field when the fiber is damaged, by using old spider silk or drawing a new supply from living spiders.

Since World War I there has been a slackening in the use of this material. The finest threads are useless for cross hairs because of their fragility and the difficulty of installation. Because dragline silk is most often used, the joined fibers must first be separated so that the primary line will be a single uniform thread. This can be easily done, since the two or four threads are discrete, and the resultant single strand, averaging 1/20,000 of an inch in diameter, is usable. Even finer fibers can sometimes be used. But the lines spun by spiderlings and small spiders, as well as the finer fibers of larger spiders, are usually quite useless.

The cocoon silk of the several large *Argiope* can often be employed for telescopes. The floss beneath the tough outer covering is pulled out easily, and single strands of considerable length procured. This cocoon silk is spun from different glands and is not quite as strong as the dragline silk, which is the most commonly used fiber. The silks of many spiders are suitable for reticules. In Europe the favorite species are large orb weavers such as *Araneus diadematus* and *Zygiella atrica.* Many other spiders provide suitable silk, even those belonging to quite different families. In the United States most silk comes from the common house orb weavers, *Nuctenea cornuta* and *sclopetaria*, from the numerous humped or oval-bodied *Araneus* and other argiopids, notably the large *Argiope aurantia.* The silk of the black widow has also been used extensively.

Silk is usually reeled from the spinnerets of living spiders and placed upon suitable frames for storage. It is easy to secure and retains its properties for many years. During World War II there was an increased demand for spider fiber for laboratory and surveying instruments. Although few of the optical instruments requiring spider silk were directly concerned with war in the field, some newspaper publicity gave the impression that the silk was in great demand as a critical war material. The truth of the matter is that all needs were filled by a few individuals who only devoted part of their time to the securing of the web.

The importance of spider silk in industry has decreased progressively during the past thirty years. Its place has been taken by platinum filaments and by engraving on glass plates. Where an aerial reticule is desired, drawn filaments of silver-coated platinum wire are frequently used. These filaments, usually 1/10,000 of an inch in diameter, are mounted in a heavy metal ring to form the desired pattern. They are said to be superior to the spider web since they show an even black line and do not sag in a humid atmosphere. For all instruments requiring a complicated pattern, etched glass reticules are usually used. In bomb sights, range finders, periscopes, and most gun sights, in fact in virtually all optical fire-control instruments, the width of the line has to be carefully adapted to the optical purposes and characteristics of the instrument. Etched glass is obviously necessary in most such instances; it would be impossible to accomplish the desired results with spider silk.

SILK FOR TEXTILES

It has for centuries been the ardent desire of araneologists to find some way of exploiting for commercial purposes the tremendous supply of spider silk available in nature. As long ago as 1709, a Frenchman, Bon de Saint-Hilaire, demonstrated that spider silk was usable for fabrics in the same way as the silk of the silkworm. A large number of egg sacs were washed, boiled, and cleansed of all extraneous matter, then allowed to dry out. With fine combs the sacs were carded and worked into slender thread of a pleasing gray color. Two or three pairs of stockings and gloves were made from the natural silk, and were presented to the French Academy. So sensational was this accomplishment that in 1710 the Academy of Sciences of Paris commissioned R. A. de Réaumur to investigate the possibility of an extensive utilization of spider silk. After a thorough study, this eminent entomologist (and inventor) concluded that there was little likelihood that spider silk, at least such as was available in Europe, could become a profitable industry.

The difficulties he enumerated are inherent in the spiders themselves and in their silk, and these difficulties have continued to rule out the silk of spiders as a potential material for commerce. In the first place, spiders are solitary, predaceous animals that feed only on living invertebrates. Each spider must be segregated and maintained apart from its neighbor; cannibalism is the rule when spiders come together, and the population is soon decimated. Space requirements are considerable and the difficulties of providing suitable food almost insurmountable. Only the egg sac silk was considered at that time to be usable and, although many sacs are produced by the females, it would require, as de Réaumur estimated, 663,522 spiders to produce a pound of silk. On such terms, competition with the silkworm was impossible.

The silk itself was considered to be inferior in strength to that of the silkworm, owing to its far finer threads which lacked the luster of insect silk and were difficult to work satisfactorily. The silkworm produces a single line of silk, usually between 400 and 700 yards long—a production representing the total output, the whole lifework, of the larval moth. Even with the relatively thick lines of the silkworm, their joining together to form commercially usable threads is an exacting process which, because no mechanical solutions have been successful, must be done by hand. Strands of spider silk do vary in thickness, and the large silk spiders of the genus *Nephila*, which abound in the East Indies, produce a silk noted for its strength. However, statements that the lines in the webs of *Nephila* sometimes attain the thickness of darning wool are exaggerations. Their thickest line is very much finer than that of the silkworm.

In Madagascar an attempt was made to take silk from the local spiders by drawing it directly from their bodies. The natives brought the animals into cleared areas and established them in great numbers near the site of the reeling apparatus. At intervals, the mature spiders were removed from their webs and imprisoned in a most curious device consisting of little stocks that held them firmly between cephalothorax and abdomen. Then small revolving mills were touched to each spinneret, and as the filaments were pulled out, they were rolled into a single thread by a hand-operated mill. The silk so produced was of a beautiful golden color and quite as good as that of the silkworm, but the project had to be abandoned because of the practical difficulties.

In the United States, Dr. B. G. Wilder drew attention to the possibility of using the silk of our big American *Nephila*. In 1866 he extracted silk directly from the body of the spider—unaware, at that time, of the earlier European experiments. Wilder was amazed by the ease with which it was possible to reel off the silk, and intrigued by the possibility of producing quantities of it for textiles. From one spider he reeled off silk for an hour and a quarter, at the rate of 6 feet per minute, taking a total of 150 yards. Later he devised an ingenious little apparatus to hold the spider during the reeling, and was able to obtain quickly the full quota of available silk. In addition to holding the creature firmly in stocks, the device had a round piece of cork on which the spider could rest its legs, thus being prevented from interfering with the flow of silk from its spinnerets.

Dr. Wilder found that one female would yield at successive reelings one grain of silk, and that 415 spiders would be required to yield one square yard of commercial silk. For an ordinary dress requiring 12 yards of material, therefore, nearly 5,000 spiders would be required. This was quite bountiful production for spiders, yet it is still only half the amount obtainable from an equal number of silkworms.

Today we are no nearer than Saint-Hilaire and Wilder to a realization of spider silk as a practical commercial textile. The basic obstacles remain, inherent in the characteristic differences between the silk spider and the silkworm.

USE OF SILK BY PRIMITIVE PEOPLES

A material of such abundance and strength as spider silk could scarcely have failed to be used by primitive peoples for some of their needs. Indeed, it is surprising that we do not have more records of its use in the Americas, where the same types of spiders abound that have supplied the Papuan and Oriental natives for generations. From the great *Nephila* spiders comes silk to supply certain New Guinea natives with gill nets, kite nets, dip nets, and various lures for their fishing activities, silk with which to weave bags, caps, and headdresses, and silk for other purposes. Strength resides not in a single strand of silk but rather in the twisted and matted threads, which form a tough fabric. The large aerial webs of *Nephila* are made with a very strong silk, and these webs are capable on occasion of ensnaring birds in their viscid and elastic lines.

In the New Hebrides, the natives use silk to fabricate small bags in which they carry arrowheads, tobacco, and even the dried poison used on their arrowheads. Some New Guinea natives of the Aroa River district make a headdress of insect or spider silk to keep out the rain. To more sinister uses are put the smothering cap and the dooming bag, both made by the New Hebrideans. The former is a strong, conical cap which is pulled down tightly over the heads of victims, usually adultresses, and causes death by suffocation. The dooming bag, a purse filled with various bric-a-brac, is said to have magical properties. According to the stories, it is rubbed over the forehead of a sleeping victim with a rhythmic motion and with muttered magical words, causing him to remain in a deep hypnotic sleep from which there is no awakening. The soporific effect of the dooming bag is assured by the victim's executioners, who administer a *coup de grâce* after they have carried him into the jungle.

Of more interest are the fishing nets of the Papuans, which show varied and ingenious use of spider fiber. Several accounts illustrating primitive man's ability to seize upon common materials and suit them to his purposes are well worth mentioning.

The North Queensland black boy entangles one end of a thin switch in the web of *Nephila* and, by adroit weaving motions, twists the coarse lines into a strand a foot or more long. The frayed ends of the line are moistened in the crushed body of the large olive-green silk spider (known to these aborigines as "karan-jamara") and the remaining morsels are thrown into the stream, immediately attracting shoals of small fish. As the silken lure is trailed through the shallow water, a fish rises to sample the tidbits on the invisible strand. Lines of gossamer become entangled in its teeth, and the smiling angler lands the two-inch long prize with a careless flourish. This method of fly fishing, and other engaging fishing techniques of the Australian aborigines, may be found described in detail in E. J. Banfield's book *Tropic Days.*

That the catch is limited to small fish does not detract from the efficiency of the method. Many fish are caught in a relatively short time, seventeen fingerlings in ten minutes according to one account, and they make up in numbers what they lack in size. It is said that these lures, as generally made, are capable of holding fish weighing nearly three-quarters of a pound.

A similar lure is used as part of a novel method of catching fish on the east coast and adjacent islands of New Guinea, and in the Solomon Islands. The natives make a kite of the large flat leaves of one of the local trees, sewing them together and stiffening them with tough strips to produce an object about two and a half feet long and nearly a foot in width. The completed kite is embellished with five wings of pandanus leaf. A flying line is made of fiber twine, ordinarily about one-third of a mile long, while the tail is another length of twine about 100 to 300 yards in length, at the end of which is tied a tassel made from the web of silk spiders. The kites are then flown over the seas either from the shore or from canoes in such a way that the spider tassel skips along the water and entices fish to strike. The golden-yellow silk entangles the teeth of the fish and, after some maneuvering with kite and boat, the fish is lifted into the canoe by means of a dip net.

Still another intriguing method of capturing small fish is practiced by certain Solomon Island natives. This account by H. B. Guppy is from *The Solomon Islands and Their Natives:*

> The following ingenious snare was employed on one occasion by my natives in Treasury, when I was anxious to obtain for Dr. Gunther some small fish that frequented the streams on the north side of the island. I was very desirous to have some of these fish, and my natives were equally anxious to display their ingenuity in catching them. They first bent a pliant switch into an oval hoop about a foot in length, over which they spread a covering of stout spider-web which was found in a wood hard by. Having placed the hoop on the surface of the water, buoying it up with two light sticks, they shook over it a portion of a nest of ants, which formed a large kind of tumour on the trunk of a neighboring tree, thus covering the web with a number of struggling young insects. This snare was allowed to float down the stream,

when the little fish, which were between two and three inches long, commenced jumping up at the white bodies of the ants from underneath the hoop, apparently not seeing the intervening web on which they lay, as it appeared nearly transparent in the water. In a short time, one of the small fish succeeded in getting its snout and gills entangled in the web, when a native at once waded in, and placing his hand under the entangled fish, secured the prize. With two or three of these web hoops we caught nine or ten of these little fish in a quarter of an hour.

The Papuan natives make landing nets from the orb webs of *Nephila*. A. E. Pratt describes this practice as follows:

One of the curiosities of Waley (near Yule Bay), and indeed, one of the greatest curiosities that I noted during my stay in New Guinea, was the spiders' web fishing-net.

In the forest at this point huge spiders' webs, 6 feet in diameter, abounded. These are woven in a large mesh, varying from 1 inch square at the outside of the web to about 1/8 inch at the centre. The web was most substantial, and had great resisting power, a fact of which the natives were not slow to avail themselves, for they have pressed into the service of man this spider, which is about the size of a small hazel-nut, with hairy, dark-brown legs, spreading to about 2 inches. This diligent creature they have beguiled into weaving their fishing-nets. At the place where the webs are thickest they set up long bamboos, bent over into a loop at the end. In a very short time the spider weaves a web on this most convenient frame, and the Papuan has his fishing-net ready to his hand. He goes down to the stream and uses it with great dexterity to catch fish of about 1 lb. weight, neither the water nor the fish sufficing to break the mesh. The usual practice is to stand on a rock in a backwater, where there is an eddy. There they watch for a fish and then dexterously dip it up and throw it on the land. Several men would set up bamboos so as to have the nets ready all together, and would then arrange little fishing parties. It seems to me that the web resisted water as readily as a duck's back.

Although Pratt's account has not been verified, there is nevertheless more reason to believe that it could be true of *Nephila* webs rather than the garden variety of orb web. It is not difficult to persuade the spider to use a bamboo hoop, since it is a most suitable framework for a web, and we know that American orb weavers sometimes oblige by spinning a web on a frame supplied to them. It is also true that the radii of the *Nephila* webs are most numerous, and that the many closely set spirals would contribute to the strength of the web. The spiral line becomes a permanent part of the web and thus multiplies its strength. Finally, it is possible that immersion in water contributes in a mechan-

maleness or maleness, and nothing in their early-life activities of digging, hunting, or web spinning that marks either sex. Many people think of immature spiders as being female, and there is good reason for this inasmuch as they usually more closely approximate the mature female in general appearance. It is probable that ancestral spiders exhibited little sexual dimorphism, and we note that this is true for some (but by no means all) of the more primitive types. Changes in the sexes have occurred both in the female and male, but they have been far greater in the male.

The female is specialized for a particular function and, if we presume to evaluate the sexes in finite terms, is a far greater contributor to the race than the male. Whereas the male has completed his assignment when he transfers the sperms to the female receptacles, the female maintains the eggs in her body until they are ready to be delivered and fertilized, encases them in a silken sac, guards them in various ways, and often is on hand to protect the young spiderlings for a considerable period. Her body has been molded as a receptacle for nurturing a variable number of developing eggs, and it responds to this need by maintaining a greater size than the male. Perhaps in response to her protective role, she is less conspicuously colored and far less of an extrovert than her male. On the other hand, because of greater size, she is much more powerful; and she is dominated most of the time by predatory instinct intensified by her solitary habits.

Among spiders, the male is a luxury item, developed for the single purpose of transferring the sperm. He offers no protection to the female or the offspring, as do many other animals, and is usually dead before the eggs are laid. He has changed in various ways to become a specialist, and is modified in many ways to play his part more expertly. The force that sends him into the arms of the female ogre is a very strong one, but he has become conditioned to preserve himself by taking flight should he be unwelcome. He has also become conditioned to overpower the female on certain occasions.

The specialization of the male has proceeded in several directions, and we find a considerable variety of types. In many of the hunting spiders the sexes are quite similar in size and seemingly nearly equal in strength. But even with these spiders there are notable differences. The abdomen of the male is slimmer, and frequently clothed with somewhat different hairs and patches of setae. The color pattern of males is almost always somewhat brighter, even though the species may be classified as drab. In these spiders of nearly equal size (the *Lycosidae, Oxyopidae, Gnaphosidae, Clubionidae*, and others), the outstanding feature of the male is his somewhat longer legs, which give him a greater range of sensory perception and are thus important in evading and overpowering the female. This disparity in leg length is presumably maintained because of and correlative to the often quite different modes of life of the sexes, and the dedication of the whole adult life of the male to sex. Among the spiders that have quite similar sexes except for the longer legs of the males are the trap-door spiders. We can interpret in various ways the difference in leg length. In addition to the advantages enjoyed during the courtship and mating, the difference in leg length may mean that the male is better fitted to wander about in search of the female. On the other hand, the longer legs may represent the more generalized condition, and the shortening of the legs of the female a response to the burrowing habit.

Sexual dimorphism mainfests itself in pronounced difference in size in many of the higher web spiders. Among the orb weavers (Plate XIX) exist all intergrades between a near size equality of the sexes to the reduction of the male size to an infinitesimal portion of the female bulk. Many large humped orb weavers have males that are nearly as large as the females, but in other members the males may be one-fourth or even less the size of the females. In the Southwest *Araneus pima*, where the females often attain 20 mm in length, the males have been measured up to half that length. The related *Araneus illaudatus* (Plate 15), typically larger than *pima*, and even sometimes attaining 30 mm in length, has pygmy males that rarely exceed 4 mm in length. The size disparity is also great in *Argiope*, *Cyclosa*, and many other genera, the male in the former genus being about one-fourth as long as that of the female. The disparity is even greater in *Nephila*, where the female of the American species weighs more than 100 times as much as the male, and in some exotic species it is said to be over 1,000 times larger than the male. The male is also a pygmy among such spiders as *Mastophora*, *Gasteracantha*, and *Micrathena*. A remarkable sexual dimorphism also exists among the comb-footed spiders, the *Theridiidae*, and the vagrant crab spiders of the family *Thomisidae*.

The smaller size of the male gives it certain advantages during courtship and mating, and perhaps is used to counterbalance the physical superiority of the female. In *Mastophora* (Plate XVIII) and *Nephila* the smaller size of the male has been carried to a ridiculous extreme. These tiny males are virtually immune to the attacks of the great females, being far beneath the usual size of the latter's prey. Tiny insects have much the same immunity and are even tolerated when they crawl over a spider's body and are left untouched when they are caught in the web. Great reduction in size doubtless represents an orthogenetic development that has nothing to do with the needs of the sex itself, but persists once it has started. It also brings with it other problems, since the males become sexually mature weeks in advance of the females and must live until the females mature.

The males possess the two pedipalps with the sexual organs at the end and these appendages may be further embellished with accessory spines or processes. The legs are also frequently armed with spurs or with rows of modified and enlarged spines that aid in clasping the female or in holding her chelicerae or appendages. The tarantulas, trap-door spiders, and many of the primitive true spiders have prominent processes on the front legs to catch the appendages of the female. The elongated chelicerae of *Tetragnatha* and *Pachygnatha* are used to grasp those of the female. Among aerial sheet weavers, the *Linyphiidae*, we find a large group of species in which the heads of the males are specialized in divers peculiar ways. Some have pointed or rounded spurs armed with curious setae, great rounded lobes, thin processes, prolongations of the clypeus or front, and the eyes are often carried to the tops or sides of these eminences. There is little doubt that these many spurs are of significance in the mating of the species. In some of these spiders, it is known that the female fixes her chelicerae in the pits on each side of the head lobe, and thus orients the male for pairing.

Sexual dimorphism also mainfests itself in profound differences in color pattern and intensity. The carmine legs and shining black body of *Atypus bicolor* (Plate 6), our largest purse-web spider, far outshine the pleasant tan or brown tones of the female. Male tarantulas have a darker body and often have the abdomen set with long golden or reddish hairs. Among the true spiders, the males are much more varied and usually more handsome than their mates. This is especially true of the jumping spiders which, especially in the tropics, display a spectrum of color, the most brilliant hues of which are restricted to the males. In many instances the sexes are so different in appearance that they were formerly regarded as being distinct species. Among certain of the sedentary spiders the sexes are somewhat more equal from the color standpoint; and of the spiny-bodied spiders, *Gasteracantha* (Plate 16) and *Microthena*, even the females have beautifully painted and sculptured bodies.

SPERM INDUCTION

The strange process by which the male spider transfers semen from the primary genital organs into the receptacles of the palpi is called "sperm induction." It was observed for the first time in 1843, by Anton Menge, who described how the male constructed a little web of silk (see Fig. 3, D), deposited a droplet of sperm upon it, and then applied his palpi to the drop until it was entirely absorbed into these latter organs. It is not at all surprising that this extraordinary action was doubted at first by many people, among them several eminent arachnologists, who insisted that there must be an internal connection between the testes, deep in the abdomen, and the tips of the palpi. Now we recognize sperm induction as only the first step in a series of strange acts that mark the sexual life of spiders. In the tiny sperm web we discern a structure analogous to the spermatophore of many arachnids.

Sperm induction is of necessity a very common phenomenon, but one must be on hand at the right time to observe it. Soon after the male becomes sexually mature, he charges his palpi and is then ready to wander about in search of a mate. This is an act that is not part of the previous experience of the male, but is initiated by internal changes in the body associated with the arrival of maturity. He performs this act instinctively and perfectly at the outset, because it is fixed in his behavior as a racial memory. Thereafter, he fills his palpi frequently, usually immediately following copulation, which is the best time to see this interesting spectacle. A few spiders are able to mate more than once without exhausting their semen; others have to pause during their mating to refill the bulbs.

There is considerable diversity in the manner in which different types of spiders accomplish their sperm induction. In no known instance is the sperm taken directly from the genital opening at the base of the abdomen, which would appear to be a logical means of solving the problem, and would be physically possible in many spiders with long palpi. Some spiders spin very simple, loose webs and absorb the semen by placing the palpi directly against it. In *Pholcus* a single silk line between the third legs is drawn across the genital opening until the spermatic globule adheres to it, whereupon it is taken up by the chelicerae and held there for direct absorption by the palpi. Some of the other primitive spiders do essentially the same thing, but hold the globule and the tiny web between the palpi or front legs until it is absorbed. A great many spiders spin a tiny sheet of very fine web, usually quadrangular or triangular in outline, place a drop upon the surface, and then take the sperm indirectly by applying their palpi on the opposite side of the sheet.

Among the tarantulas sperm induction is a long operation that sometimes requires three or four hours. The male spins a large flat sheet of silk, attached to adjacent objects, in which are left a large oval opening and a much smaller one, the two separated by a narrow band of strongly woven silk. He then crawls through the large opening and, lying on his back, strengthens the silk around the holes. The area in front of the genital opening is rubbed against the reinforced silk band to lay down additional silk lines from glands in that area. Then after rubbing his palpi between his chelicerae and further stroking of his genital area a drop of spermatic fluid appears and is deposited on the underside. The male now clambers back and, sitting upright on the web over the globule, reaches around the edge of the narrow band to touch the sperm directly. The process of absorption takes an hour or more, and consists of a rhythmical alternate tapping of the palpi in the globule, usually at the fast rate of 100 to 150 taps per minute. Afterward the web is destroyed or deserted.

Most spiders are able to recharge their palpi much more quickly, usually within half an hour. T. H. Montgomery has described how one of the small American wolf spiders, *Schizocosa crassipes*, spun a triangular sheet attached to the floor and walls of its cage, and stood on the upper side of the web. A small globule of yellowish semen was ejaculated upon the surface of the sheet at about the middle. The male then "reached his palpi downward and backward, below the sheet, and applied the concave portion of the palpal organ of each against that part of the sheet which carried the drop of sperm. Each palpus was then rubbed against the lower surface of this drop several times, then withdrawn and slowly shaken in the air, while the other was similarly applied to the drop." This process continued for seven minutes, during which time all the sperm was taken up into the palpal organ, and soon afterward the male left the sperm web.

The male seems to derive considerable gratification from the process of sperm induction. Before the act, the genital region is rubbed against the strands of webbing to incite the ejaculation of the semen. The presence of the female is not a necessary adjunct of the act, which is implemented by internal factors, whereas later on she, or her threads, become the stimuli which result in the mating.

COURTSHIP

Inasmuch as the young male leads the same kind of life as the female, and lives in similar webs on plants or hides in similar places on the ground, maturity finds him not far distant from female neighbors. After he has prepared himself for mating by charging his palpi, a new impulse sends him in search of a mate, and he moves about in a random manner until he is able to detect a female.

Since spiders are largely creatures of touch, it is not surprising that to find the female he relies mainly on the fine sensory hairs that clothe his body and appendages. Contact with the substratum brings him something more than the mere mechanical sensation of touch or tension or vibration. Accompanying it is an ability to distinguish certain chemical substances with which his hairs come in contact; this combined sense is called chemotactic. The receptors for it have not been recognized, but it seems certain that some of the hairs are sensillae that respond to this type of stimulation. Since the sensation comes to the spider only when in actual contact with chemical substances, the sensation is nearer to what we call taste than any other sense possessed by man. There is recent new evidence that certain presumably chemical distance messengers called "pheromones" are capable of eliciting reactions from spiders in various aspects of their lives. The male spider mostly becomes aware of the presence of a mate through the touch of her threads, or of the trail she leaves on the substratum, or of her actual body.

There is still another way in which some spiders are able to discover their mates. Some vagrant spiders have developed eyes of such acuity that they can see moving objects at a considerable distance—for a spider—and can identify the other sex when still several inches away. In these relatively longsighted types, and especially in the jumping spiders, recognition of the female may be possible by sight alone, without any aid from the chemotactic sense. On the other hand, certain of the wolf spiders, having vision nearly on a par with the jumpers, nevertheless appear to require both sight and touch to incite pairing.

Once the male has discovered the female, he is on the threshold of realizing the racial instinct for which he has become specialized—the transfer of the semen. But there are difficulties. The object of his attention may not be of the same mind as he is, and she usually exceeds him in size and strength. Further, since virtually all of her life has been devoted to capturing and feeding on animals of suitable size, her first instinct is predatory. That the interloper is a male of her own kind is immaterial if she is not conditioned to distinguish him from any other suitable prey. There consequently must ensue certain more or less marked preliminary activities before the actual mating; these preliminary activities constitute the spiders' courtship. Most of the initiative is taken by the male who—being the less valuable sex—is conditioned to make the first advances and brave the danger. Upon him rests the burden of announcing himself in a convincing manner, and of stimulating the female to a point where sexual union is possible.

Among the aerial spiders and other web spinners, courtship usually consists first of some kind of vibration of the threads of the web, and later of stroking the body of the female. Among the hunting spiders there is a considerable diversity in methods of courting. Those species blessed with good eyesight have developed a relatively complicated prenuptial procedure during the course of which the male advertises his presence by movements of the legs and body. Correlated with this behavior to some degree are various epigamic structures such as brushes or ornaments on the legs and tufts of hair on the head. Spiders with poorer eyesight are ordinarily much more conservative in their prenuptial routine, since the female would be unable to see the details; but occasionally body ornaments are present in this group as well. There are numerous intergrades between a well-marked courtship, as exemplified in the bizarre love dances of the jumping spiders, and almost no courtship at all; the fundamental mechanism and the particular path that each group has followed to arrive at its present specialization are subjects that must be discussed at length.

The prime descriptive and analytical studies of spider courtship and sexual biology, following the classical work of Anton Menge, were made in the United States by G. W. and E. G. Peckham in 1889 and 1890, and by T. H. Montgomery in 1903 and 1910. In addition to fascinating descriptions of the sexual processes in many species, quite adequate explanations of the significance and evolution of various phenomena in terms of selection were presented. The Peckhams were strong exponents of sexual selection as outlined by Darwin, and concluded that the female jumping spider responded to the charm and beauty of the posturing male and made a conscious selection of a mate. They believed that the males were more numerous and, especially in cases where there was male dimorphism, that the brighter male was preferred by the female. They argued that the numerous ornamental features on the bodies of the male jumping spiders were developed as a result of sexual selection. They rejected A. R. Wallace's views that such epigamic characters were a result of a surplus of vital energy that went with maleness; because the male was more vigorous, he was more highly colored and likely to be more successful in his suit with the female, and this would more surely and more often leave progeny.

In 1910 Montgomery rose to the defense of ordinary natural selection, and in a masterful essay virtually refuted the claims of the Peckhams with regard to sexual selection. Montgomery believed that the adult male "is excited simultaneously by fear of and desire for the female, and his courting motions are for the most part exaggerations of ordinary motions of fear and timidity. By such motions he advertises himself to the female as a male, but there is no proof that he consciously seeks to arouse her eagerness by esthetic display . . . there seems to be no good reason to hold that the female is actuated in her choice by sensations of beauty." Montgomery defined courtship in a more limited way than do modern arachnologists, and believed that in some vagrants there was no courtship at all. However, judging from his descriptions, his interpretation is in most cases a modern one. Commenting upon spiders that have good sight, he said as follows: "What we do know is that the male by his courtship, a set of motions resulting from the conflicting states of sexual desire and fear, exhibits or advertises himself as a male; and that the female on sight of this courtship recognizes him as a male and accepts him if she be eager, or else becomes gradually stimulated by watching him." Montgomery further believed that many secondary sexual characters in the male "may be most readily explained as being conserved by simple selection. Peculiar male ornamentation would be selected because it insured quicker sex-recognition, therefore prompter mating. The male is thereby more surely accepted by the female, not selected by her in the sense of Darwin. The process is much more an announcement of sex by the male than a choice by the female, and results in the female accepting the sex rather than the individual." Montgomery did not subscribe to Wallace's belief that the males exhibited a higher degree of vitality, but argued instead that the need of greater protection by the females was the reason for their less conspicuous coloration, as in birds.

It has remained for W. S. Bristowe to take up the problem where Montgomery left off, and to extend and elaborate his thesis on the basis of a much vaster literature and innumerable observations on European spiders. Bristowe's views were presented in convincing fashion in 1929, in a long paper entitled "The Mating Habits of Spiders, with special reference to the Problems surrounding Sex Dimorphism." In this treatise he pointed out that the complicated visual displays of the jumping spiders probably arose by ordinary natural selection.

Primitive spiders were shortsighted hunters that groped their way as they walked and stretched out their front legs to test the substratum. Perception of the environment was accomplished by a chemotactic sense largely confined to the extremities of the appendages. Since sight was limited, it was necessary for the male to touch the spoor, the threads, or the body of the female to discover her presence. Since the males were able to detect the presence of a mate often before she was touched, those males that started to advertise their identity early by means of their front legs were more likely to survive the assault of their larger, predaceous mates. Movement of the appendages and parts of the body enhanced the chances of survival and also increased the possibility of finding a mate. All these advances in posturing were accompanied by a gradual improvement in the acuity of the eyes, likewise arrived at by selection. Males tend to produce more pigment than females, so those males that were able to develop strikingly colored spots in front that were visible to the female, were able to survive more often. The various antics and decorations worked hand in hand.

The most generalized types of courtship are exhibited by those spiders in which distance perception is feebly developed, the majority being *shortsighted hunters.* More specialized displays have arisen in two ways: by improvement in the acuity of the eyes, as in the *longsighted hunters;* and by development of expansive webs that enlarge the limits of perception by touch, as in the *web builders.* These divisions approximate in a general way the systematic position of the species.

The Shortsighted Hunters. Most of these spiders are nocturnal creatures of the ground that rely almost entirely on their chemotactic sense of touch to inform them of the character of their environment. They test the surface by means of their legs, which act both as organs of touch and offensive weapons. Their approach to the female is usually a bold one, since most of them approximate her in size, and a mere touch is sufficient to inform them of her sex and species. Further, this physical contact probably gives the female the same information. Recognition is almost instantaneous and largely based on the chemical sense. There remains only for the male to stimulate the female until mating can be accomplished. He does this by stroking and tickling her body, while at the same time maintaining a firm grip on her with his legs or chelicerae, so that she cannot escape.

The tarantulas are wonderful subjects for the study of mating behavior in the shortsighted spiders. Alexander Petrunkevitch has described the courtship and mating of *Dugesiella hentzi* thus:

> When the restlessly wandering male happens to touch with his legs some part of the body or a leg of the female, he at once stops short and begins to strike simultaneously and violently with his anterior, sometimes with all four front feet. . . . This continuous beating with the front legs upon the body or legs of the female constitutes the first step in the courtship on the part of the male. In case the female does not attempt to run away, the male soon shifts

his position until he is facing the female. The behavior of the female during the first stage of the courtship is composed of two elements. At the first touch she raises the front legs and assumes the attitude of defense and threat. The subsequent touching results in her rising high on her hind legs while still holding up the front legs. Finally, she opens the fangs and the male catches them with the hooks on his front legs. . . . They serve admirably to guard the male against possible injury or even death while at the same time aiding him in the act of coitus. For he now forcibly pushes back the cephalothorax of the female with his front legs and drums with the patellas of the palpi on her sternum, all the time advancing.

The mating that follows lasts only a minute or two, after which the two sexes part, the female ordinarily making no attempt to attack the male.

Many of the small running spiders spin little silken cells under stones or in tiny nooks on trees. *Wulfila saltabunda*, one of the smaller anyphaenids, weaves a little curtain beneath the leaf of an herb or bush and stands upright on the silk. The male stands beneath the sheet and drums on it with his long front legs and palpi, and at intervals pulsates his abdomen up and down. The female often responds by tapping with her front legs and palpi, and vibrates the sheet immediately above the male. The male will court the female in this position for hours, but mating ordinarily does not occur until evening. He seems able to avoid the female without great effort, and to be relatively immune to her attacks. However, she is much more powerful and can kill him with ease if he approaches her too insistently when she is pregnant or otherwise not ready for mating.

Some of the gnaphosids, notably *Drassodes* and *Zelotes*, are said to take possession of an immature female by enclosing her in a silken cell. Just after her final molt and before she has attained her full strength, the male mates with her. This is possible since the male matures earlier than the female and is able to recognize her as a prospective mate even though she is immature. It is the habit of many of these spiders to live in adjacent silken sacs under the same stone or piece of bark. Not uncommonly, a male in the penultimate stage, when he presumably has no instinct for recognizing or sequestering a mate, will be found in a sac with an immature female. This suggests that the association in many instances may be only a fortuitous one.

A few of the sedentary spiders with inferior eyesight have given up life on webs and have become vagrant secondarily. An interesting example is that of *Pachygnatha*, one of the big-jawed spiders that live in grass and vegetation especially in cattail marshes. The male prowls among the grass roots and finds his mate by touch. He seizes her and, aided by special spines and long teeth on his chelicerae, holds hers firmly until her mating instincts have become aroused or her hostility forces his retreat.

Among the misumenoid crab spiders we find few of the preliminary activities identifiable as true courtship. These stout spiders live on the ground and on plants and hunt as well by day as night. The eyes of some are fairly large, but the crab spiders seem to make little use of sight in their hunting or courting, a fact which may be partially accounted for by their habits of deception and inactivity. When a male discovers the female of his species, he immediately climbs upon her back or seizes an appendage with his chelicerae. He is much more agile, and can tickle and caress her body until he is able to accomplish his purpose. In some of these spiders the males are very much smaller, and usually more darkly marked than the females. The latter often walk around with a tiny long-legged male clinging to their backs, paying little attention to his activities.

Certain spiders have interpolated in the sequence of their courtship a habitual act that tends to set them apart from all others. The males of the stocky little species of *Xysticus*, preparatory to the mating, spin thin webs over the female, attaching tiny silken lines from her abdomen and legs to the substratum. This web has been called the bridal veil, and its spinning is one of the extraordinary prenuptial habits of many crab spiders.

The Longsighted Hunters. It is among the spiders of this group that we find those notorious for their love dances (Fig. 2). Almost all are day hunters, a habit in keeping with their need for light to display themselves properly during courtship. Some are well-known vagrants, and have received such expressive names as "wolf spider," "lynx spider," and "jumping spider" in recognition of their life of action. However, even in this group with the best eyes, reliance is only partially placed on sight during courtship; and in most instances the event does not occur unless the male actually touches the female, even though he may perceive her by sight.

The whole makeup of the prenuptial display—posture, antics, and epigamic ornaments—is distinct for each species. While developing these features nature has had to keep many allied species separate, and thus has evolved by selection many different kinds of dances. The female has become conditioned to respond only to those dances performed by her species, and she rarely makes mistakes. The actual mating is usually preceded by a certain amount of stimulation by the legs of the male, and it is this final action that completely precludes the possibility of any related species being accepted.

In most wolf spiders the palpi and front legs are provided with some kind of ornamentation that contrasts sharply with the rest of the body. When such epigamic characters are present, display of them is usually part of the courtship ritual. Among the small *Pardosa* the palpi, usually bedecked with jet-black hairs or variegated with black and snow-white ones, receive a large share of the orna-

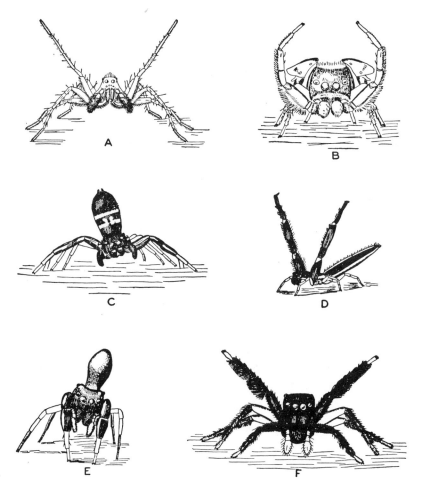

Fig. 2. Courtship postures of male wolf and jumping spiders. A. *Pardosa milvina.* B. *Pellenes viridipes.* C. *Peckhamia noxiosa.* D. *Marpissa pikei.* E. *Peckhamia picata.* F. *Euophrys monadnock.* (Redrawn from Kaston, Emerton, and the Peckhams.)

mentation. The larger lycosids usually have the front legs darkened and occasionally provided with brushes of conspicuous black hairs.

The courtship patterns of American wolf spiders were first investigated in detail by T. H. Montgomery in 1910, were the subject of a special analysis by B. J. Kaston in 1936, and during recent years have been intensively studied by J. S. Rovner from the basis of mechanisms controlling the agonistic display and sexual behavior. Acoustic communication by stridulatory devices on the palpi of males, that produce sounds audible to the female, has now been verified for many lycosine species and thought especially important during night courtship. It has also been suggested that pheromones play an important role in the lives of lycosid and other spiders, but evidence for airborne, chemical messengers is still not clearly established.

Only a few of our species can be mentioned. *Pardosa milvina* (Plate XXVI) is a small, long-legged wolf spider with black head and palpi. The following description is from Montgomery:

> The courtship motions are as follows: The male stands with his body well elevated above the ground (an attitude that a female takes only when she is aggressive) on his three posterior pairs of legs, his head higher than his abdomen, so that the long axis of his body describes an angle of 30°–40° with the surface of the ground. He waves his palpi upward in the air (i.e., straightening them out before his head) flexes them outward, from one to three times, then draws his body slightly backward and downward, rapidly waving in the air the outstretched palpi and first pair of legs, and spasmodically shaking the whole body with the violence of the movement. The vehemence and to some extent the attitudes reminds one forcibly of a small terrier barking at a cat. The movement of the palpi exhibits most clearly their relatively huge, black terminal joints. When the male is timid, or not very eager, he may wave only his palpi, and these slowly and alternately instead of together. The male repeats these motions several times, usually becoming more vehement each time, then moves a step nearer the female, repeats them again, moves nearer again, so that in a short time his outstretched shaking forelegs come in contact with the female.

A closely related species, *Pardosa saxatilis*, raises the forelegs alternately and at the same time wigwags with his jet-black palpi, using them alternately as well. In *Pardosa emertoni*, the front legs are held up in the air and the palpi are flexed and jerked, and followed by movements of the abdomen. *Pardosa modica* makes little use of his front legs during the visual display but wigwags with his palpi, often standing high on the tips of his tarsi. Within the same group such striking differences in courtship are often found.

Lycosa gulosa is a common grassland spider varying in color from gray to nearly black, and exhibiting only slight differences between the sexes. From its courtship antics it once was given the common name of "purring spider" because of the tapping of the palps against dry leaves. B. J. Kaston explains:

> Immediately upon coming in contact with the female, or within three minutes thereof, the male begins to drum his palps rapidly against the floor of the cage. These drumming movements are made so rapidly that a distinct purring or humming sound can be heard. The palps are used alternately and are raised

only a very short distance during the process. The body is held at an angle so that the posterior end of the abdomen almost touches the floor. As a consequence when the male begins to twitch his abdomen in a vertical plane, the tip strikes the floor. However, I could not detect any sounds made by this part of the body. It is highly probable that the vibrations set up in the substratum by the tapping movements of the palps and abdomen are perceived by the female. This may exert an exciting influence on her in a manner analogous to that which occurs in web-building species, where the male tweaks the threads of the female's snare.

The male of *Schizocosa crassipes* has a thick covering of black hairs on the tibiae of the front pair of legs, which are conspicuous epigamic brushes. He extends his long first legs out in front and taps the floor with both about four or five times, simultaneously and in rapid succession. Then the forelegs are raised and the body is elevated high over the posterior legs, while at the same time the palpi are extended downward to touch the floor below the face. In this position, the brushes on the front tibiae are conspicuously displayed. He advances toward the female with a rhythmically repeated waving of legs, jerking of body, and posturing. A closely allied species, *Schizocosa bilineata*, bedecked with similar ornaments on the front legs, seems on the other hand to make no use of them during mating; in fact, seems to have no visual courtship at all.

The European *Pisaura mirabilis* is remarkable for its habit of presenting the female with a fly as an inducement to mating. W. S. Bristowe has described the activity in the following manner:

> A male was given a fly and placed in a box with a female. He proceeded to enwrap the fly with silk, and then walked about with it in a jerky fashion until presently the attention of the female was attracted, and she approached him. He held out the fly to her, and after testing it with her falces, she seized hold of it. The male then crept to a position almost underneath the female, a little to one side, and inserted his right palp. After twenty-five minutes he withdrew his palp and joined the female at the fly. This is a rather remarkable piece of instinct—a carnivorous creature like a spider deliberately giving up his food as an offering to the female.

To George and Elizabeth Peckham, talented students and keen observers from Milwaukee, Wisconsin, belongs the credit for much that we know about the sexual biology of our jumping spiders. It was they who first brought to the attention of naturalists the bizarre courtship antics of the American jumping spiders. The females of this group are for the most part pleasantly colored in grays and browns, while upon the males has been showered an infinite variety of color and ornamentation. The chelicerae are enlarged, molded into odd form, and usually colored in iridescent purple, green, or gold. The principal feature of the face is a row of four great pearly-white eyes, and it is embellished above with crests or plumes and overhung with bright hairy fringes. The first legs are wonderfully ornamented with peculiar enlargements of striking colors, and often clothed with fringes of long colored hairs, pendant scales, and enlarged spines. Although less attention has been given to the other legs, they also are sometimes supplied with unusual ornamentation.

The Peckhams thought that the male jumpers were much more numerous and that "it was highly improbable that a female ever mates with the first male that comes along. . . . She rejects the advances of one after another; she flies and is pursued; she watches, with great attention, the display of many males, turning her head from side to side as they move back and forth before her; she becomes so charmed as even to respond with motions of her own body. If we can judge by her attitude, she is observant of every posture that the male takes, and appreciative of his every claim of beauty." Whereas we reject the sexual selection of the Peckhams as not truly representing the facts, it must be admitted that the final results are the same. Through the elimination of certain males directly by killing them for food, and indirectly by rejecting them as mates, there is an active female selection.

During their antics, the male jumping spiders make every effort to bring into position the striking features of their bodies. Many stretch out their front legs and wave them rhythmically and insistently, or take an imposing attitude with arms outstretched like a semaphore. Others lower these legs and keep them motionless so that nothing interferes with the view of the bands and marks on the head and clypeus. Some tilt upward to display an iridescent rose or gleaming metallic abdomen. In some the intensity of the dance verges on frenzy, whereas others perform their pantomime with grace and decorum. Among these jumpers it is also possible that airborne pheromones reinforce vision as active courtship agents. Some fascinating descriptions are given by the Peckhams.

Tutelina elegans (Plate 26) is one of the most common eastern American jumping spiders.

Both sexes are beautiful. The male is covered with iridescent scales, his general color being green; in the female the coloring is dark, but iridescent, and in certain lights has lovely rosy tints. In the sunlight both shine with the metallic splendor of humming-birds. The male alone has a superciliary fringe of hairs on either side of his head, his first legs being also longer and more adorned than those of his mate. The female is much larger, and her loveliness is accompanied by an extreme irritability of temper which the male seems to regard as a constant menace to his safety, but his eagerness being great, and his manners devoted and tender, he gradually overcomes her opposition. Her change of mood is only brought about after much patient courting on his part. While from three to five inches distant from her, he begins to wave his plumy first legs in a way that reminds one of a windmill. She eyes him fiercely and he keeps at a proper distance for a long time. If he comes close, she dashes at him and he quickly retreats. Sometimes he becomes bolder and when within an inch, pauses, with the first legs outstretched before him, not raised as is common in other species; the palpi are also held stiffly out in front with the points together. Again she drives him off, and so the play continues. Now the male grows excited as he approaches her, and while still several inches away, whirls completely around and around; pausing, he runs closer and begins to make his abdomen quiver as he stands on tiptoe in front of her. Prancing from side to side, he grows bolder and bolder, while she seems less fierce, and yielding to the excitement lifts up her magnificently iridescent abdomen, holding it at one time vertically and at another sideways to him. She no longer rushes at him, but retreats a little as he approaches. At last he comes close to her, lying flat, with his first legs stretched out and quivering. With the tips of his front legs he gently pats her; this seems to arouse the old demon of resistance, and she drives him back. Again and again he pats her with a caressing movement, gradually creeping nearer and nearer, which she now permits without resistance until he crawls over her head to her abdomen, far enough to reach the epigynum with his palpus.

The largest American jumping spiders are the massive, hairy species of *Phidippus* (Plates 27, 28), which are gaily marked with white spots and often gaudily colored in carmine, orange, and yellow. The face is usually distinguished by tufts of curled hairs and bands of colored scales and hairs. The elegance of their front legs is especially notable, with long flowing fringes of colored hairs. Some species wave these handsome legs so vigorously that they cross at the tips, but in most instances they are brought to an angle of about forty-five degrees and, as the male sways toward the female or approaches her in zigzag fashion, the legs are moved up and down to bring into view the plumes and iridescent plates.

Representative of a related group very numerous in species is *Metaphidippus protervus*. When courting, this species approaches the female rapidly until it is a couple of inches away, arms extended upward, then stops and drops the arms down close to the surface. In this position, the face, variegated with snow-white bands and with contrasting gleaming bronze scales, becomes the center of attention.

Peckhamia picata (Plate 26), one of the antlike spiders, has also received the attention of the Peckhams as follows:

The most important difference in the sexes is the greater thickening of the first legs of the male. These are flattened on the anterior surface and are of a brightly iridescent steel-blue color. Unlike most of the Attid males, this species keeps all its feet on the ground during his courtship; raising himself on the tips of the posterior six, he slightly inclines his head downward by bending his front legs, their convex surface being always turned forward. His abdomen is lifted vertically so that it is at a right angle to the plane of the cephalothorax. In this position he sways from side to side. After a moment, he drops the abdomen, runs a few steps nearer the female, and then tips his body and begins to sway again. Now he runs in one direction, now in another, pausing every few moments to rock from side to side and to bend his brilliant legs so that she may look full at them.

The little male of *Habrocestum pulex* is not so gaily colored as some of his relatives, but he makes up in enthusiasm for his lack of brilliance. His whirling dance has been excellently described by the Peckhams:

He saw her as she stood perfectly still, twelve inches away; the glance seemed to excite him and he at once moved toward her; when some four inches from her he stood still and then began the most remarkable performances that an amorous male could offer to an admiring female. She eyes him eagerly, changing her position from time to time so that he might always be in view. He, raising his whole body on one side by straightening out the legs, and lowering it on the other by folding the first two pairs of legs up and under, leaned so far over as to be in danger of losing his balance, which he only maintained by sidling rapidly toward the lowered side. The palpus, too, on this side was turned back to correspond to the direction of the legs nearest it. He moved in a semicircle for about two inches and then instantly reversed the position of his legs and circled in the opposite direction, gradually approaching nearer and nearer to the female. Now she dashes toward him, while he, raising his first pair of legs, extends them upward and forward as if to hold her off, but withal slowly retreats. Again and again he circles from side to side, she gazing toward him in a softer mood, evidently admiring the grace of his antics. This is repeated until we have counted 111 circles made by the ardent little male. Now he approaches nearer and nearer and when almost within reach, whirls madly around and around her, she joining and whirling with him in a giddy maze. Again he falls back and resumes his semicircular motions, with his body tilted over; she, all excitement, lowers her head and raises her body so that it is almost vertical; both draw nearer; she moves slowly under him, he crawling over her head, and the mating is accomplished.

Among the many American species of *Pellenes*, or *Habronattus*, the front legs and the face are lavished with decoration. The enlarged tibiae of *oregonense*, the hirsute legs of *agilis*, the iridescent blue metatarsi of *hirsutus*, the pink palpi,

Plate 1

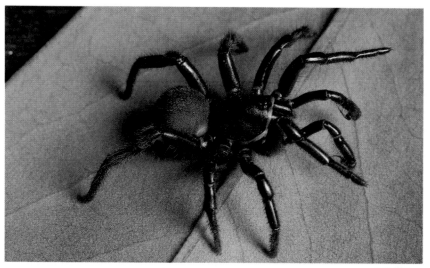

H. K. Wallace

a. *Ummidia audouini*, male

H. K. Wallace

b. *Ummidia audouini*, female

TRAP DOOR SPIDERS

Plate 2

H. K. Wallace

a. *Cyclocosmia truncata*, female, view of sclerotized disk of abdomen

H. K. Wallace

b. *Cyclocosmia truncata*, female, dorsal view

TRAP DOOR SPIDERS

Plate 3

V. and B. Roth

Seeman's tarantula, *Aphonopelma seemani*

COSTARICAN TARANTULA

Plate 4

B. J. Kaston

Orange-kneed tarantula, *Brachypelma smithi*

MEXICAN TARANTULA

Plate 5

Bernard Coyle

a. The legs are slowly withdrawn

Bernard Coyle

b. Freed spider and shed skin

MOLTING TARANTULA

Plate 6

H. K. Wallace

a. *Atypus bicolor* male on bark

H. K. Wallace

b. Purse web of *Atypus bicolor* on tree

PURSE-WEB SPIDER

Plate 7

a. Purse webs of *Atypus abboti* against tree

b. *Atypus abboti* male

c. *Atypus abboti*, female

PURSE-WEB SPIDER

Plate 8

H. K. Wallace

a. Female *Dinopus spinosus* on wall

H. K. Wallace

b. Retarius web of *Dinopus spinosus*

DINOPID STICK SPIDER

Plate 9

M. W. Tyler

a. Southern black widow, *Latrodectus mactans,* with egg sac

R. C. Kern

b. Southern black widow, *Latrodectus mactans,* in web

WIDOW SPIDERS

Plate 10

H. K. Wallace

a. Northern black widow, *Latrodectus variolus,* in web

H. K. Wallace

b. Western black widow, *Latrodectus hesperus,* on leaf

MALE WIDOW SPIDERS

Plate 11

a. Red widow, *Latrodectus bishopi*

b. Brown widow, *Latrodectus geometricus*

WIDOW SPIDERS

Plate 12

Bucky Reeves

ORANGE ARGIOPE, *Argiope aurantia*, IN WEB, DORSAL VIEW

Plate 13

B. J. Kaston

Banded Argiope, *Argiope trifasciata*, in web with swathed prey

Plate 14

R. C. Kern

a. Female on web, dorsal view

R. C. Kern

b. Female on web, ventral view

SOUTHERN ORB WEAVER, *Eriophora ravilla*

Plate 15

V. B. Scheffer

a. *Araneus trifolium*, in web

V. D. Roth

b. *Araneus illaudatus*

ORB WEAVERS

Plate 16

B. J. Kaston

a. *Verrucosa arenata*, female on egg sac

B. J. Kaston

b. *Gasteracantha cancriformis*, female

ORB WEAVERS

scarlet clypeus, banded faces of other species, are only a few of the expressions of color and ornament in the group. Some species have the third legs modified, and among them is *viridipes* (Fig. 2, B) with a strangely formed patella armed with a pale spine and marked with a black spot. During courtship he finds it a difficult task to balance himself while endeavoring to exhibit two pairs of legs. The Peckhams description follows:

> When he gets to within an inch of the female, he lifts the first legs nearly at right angles with the body, giving them a bowed position, with the tips approaching each other, so that each leg describes a semicircle, while the palpi are held firmly together in front. Up to this time he has held the body well above the ground, but now he lowers it by spreading out the second and fourth pairs, at the same time bringing the tips of the third pair nearer the body and arching the legs over the posterior part of the cephalothorax in such a way that the proximal ends of the tibiae nearly meet. As he stands in this position, the female, who is watching him eagerly, has the front surface of the apophysis plainly in view over the dorsal surface of the cephalothorax, and face and clypeus are also well exposed. Now he approaches her very slowly, with a sort of creeping movement. When almost near enough to touch her, he begins a very complicated movement with the first pair of legs. Directing them obliquely forward, he again and again rotates each leg around an imaginary point just beyond the tip; when they are at the lowest point of the circle, he suddenly snaps the tarsus and metatarsus upward, stiffening and raising the leg and thus exposing more completely its under-surface. While this is going on with the first pair, he is continually jerking the third pair up higher over his back, as though unable to get them into a satisfactory position, and the abdomen is kept twitching.

Such a display can be carried even farther to include virtually all the legs. *Euophrys monadnock* (Fig. 2, F) is a boreal spider that lives in the moss and lichens of open pine forests, frequently being found in the western mountains. The handsome little male displays the orange femora of his hind legs when he performs before the female, as told by the Peckhams.

> The palpi, jet-black with yellow ends, hung down in front; the first legs, black with pale tips, and fringed with long, thick, purplish scales, were thrown diagonally upward; the body was raised high on the tarsi of the second and third pairs, the third being lifted so that the colored femora would be seen over the second, while the legs of the fourth pair were dropped and held at just the angle that brought the femora into view between those of the second and those of the third pair. In his difficult attitude, the spider began to move. There was none of the awkwardness shown by *Pellenes* in trying to keep the

third leg in position; indeed, there was no muscular action visible as he glided smoothly back and forth, while the female, turning from side to side, kept him constantly in sight.

The Web Builders. The spiders in this category are for the most part species that have poor eyesight. Many are confirmed sedentary types or put a considerable reliance on silk, thus effectively obviating a real need for keen vision. In fact, they have extended their snares in ways that carry far beyond the limits of ordinary sight. Through the medium of her web threads, the male is able to perceive with reasonable certainty the presence of a female of his species, and to diagnose her attitude. Courtship among the web spinners usually consists of finding out how the land lies, then telegraphing to the occupant of the web the arrival of the male. In later stages a tactile stroking of the body precedes the coition. The male web spinner has an advantage in that he can approach the female at a distance and is not immediately vulnerable to her attack. A hasty retreat follows notice from the female that he is unwelcome. The whole routine of tweaking the threads, approaching the female on the surface of the web, and further stimulating her at close quarters, constitutes a tactile display of courtship equal in interest to that of the longsighted spiders.

Among the web builders we find some females that are very tolerant of their males and accept their advances eagerly. Some live together quite amicably for weeks, and others are gregarious by habit, spinning huge communal webs in which the sexes live in seeming equality. Withal, there also exist in this group females notorious for their aggressiveness, which are known to destroy their mates before or after the mating. Specialization in the orb weavers has taken them along a path where there is a premium for vigorous response to the touching of their snare by any interloper. The spider hurls herself over the web and, by a remarkable exhibition of trapeze artistry, quickly subdues and enswathes its prey. It is therefore not surprising that so finely trained an aggressor should occasionally fail to recognize the advances of a male of her species. The latter is in especial danger if the female is not fully adult, has already mated, or is gravid. After the mating the male is in great danger if he tarries too long; it is then that he is most often killed.

Among the web spinners are some that are more closely related to the longsighted hunters than to their own group and, except for the silken sheet web over which they run in an upright position, resemble the former in their courtship and mating attitudes. Males of the *Agelenidae* are in most cases equal to the females in size, and superior in agility. *Agelenopsis pennsylvanica*, a common grass spider of the eastern states, moves upon the web of the female and signals to her by tapping the silk with his legs and palpi. His advance is usually slow and measured until he is able to touch her with his legs, whereupon he actively seizes her. In most instances resistance is not strong; the male grasps her hind femora in his chelicerae and carries her to the entrance of the silken tunnel, where mating often occurs. He throws her on her side and, his head pointing in the opposite direction from hers, turns her over and applies the palpi from either side.

The tangled-web spinners (family *Theridiidae*) include many species among which the females show little hostility to the male before mating and actively enter into the preliminary maneuvers. The description of the courtship of *Achaearanea tepidariorum* by Montgomery illustrates the general habit of the whole genus.

The introductory steps of the mating are as often made by the female as by the male, and she often shows quite an insatiable eagerness, even sometimes leaving food to approach the male. As soon as the male commences to move upon her web she recognizes him as a male of her own species and, when she is eager, commences immediately to signal to him, both spiders being on the lower surface of the web and upside down (the usual position). The female hangs to the web with the third and fourth pairs of legs, and shakes the longer second and first pairs vigorously and spasmodically in the air (when those legs are not attached to web lines), otherwise with them she shakes web lines to which they are hooked. This "signalling" is a sign of eagerness on the part of the female, and so far as I have observed she makes it at no other time than when she is eager and notices the approach of a male of her own species. There are individual differences in the mode of signalling, as well as differences in accord with the degree of eagerness of the female; sometimes a female signals without moving from her original position, sometimes with the signalling she moves by short steps toward the male. When she is not eager she either remains motionless, or else rushes hostilely toward the male as an object of prey; in both cases the male makes no advances, and when she is markedly aggressive he escapes by dropping from the web. The whole attitude of the male is that of combined timidity and eagerness; he is much smaller than the female and upon a foreign web, and usually acts with great caution.

In this species the female may mate with many males and, except when heavy with eggs, rarely rejects the advances of a suitor. Whereas the male usually leaves hurriedly after mating, in some species of this group he moves to one side of the web, refills his palpus with semen, and returns to mate frequently. Among the sheet weavers of the family *Linyphiidae*, the male is also a privileged consort and is only rarely menaced by an intractable female.

The male spider must often coax the female out from her retreat before mating; sometimes he spins a series of lines as a bower in which the pairing can take place. The female of *Metepeira labyrinthea* spins a labyrinth of tangled threads behind her web and stays in it much of the time. Other orb weavers hide in a leafy retreat near their webs and communicate with the orb by means of a signal line held in the claws of one of the legs. According to G. H. Locket, the English arachnologist to whom we owe much for his keen observations of the web-spinning spiders, the male of *Zygiella x-notata* "climbs to the center of the female's web and usually seizes the line communicating with the female's hiding place with his four front legs. With his back legs he seizes one of the adjacent radii at the center and starts a series of jerking and plucking movements on the communicating line, using himself as a sort of spring at the angle of the radii. If the female does not respond he then usually climbs to her retreat, but returns again after an interplay of legs . . . eventually the female comes out, also making plucking motions, and, after a short interplay of legs, the male begins making thrusts at her epigynum; the palps are then applied alternately."

Among the orb weavers we find some species in which the male is a mere pygmy hardly worthy of the female's notice. During the mating season there are often three or four tiny males, only one-fourth the length of the female, hanging in the outskirts of the web of the large orange *Argiope aurantia*. These tiny males make known their presence by plucking and vibrating the web lines. When one of these pygmy males approaches *aurantia* from in front, she moves the web in warning and the male either moves away or drops from the web. When the male comes from behind she pays no attention to him until he creeps on her body, whereupon she slowly raises one of her legs and brushes him off. Only the persistent males are successful.

No observations have been made of the mating of the bolas spiders of the genus *Mastophora*. The female is a phlegmatic creature that lies completely immobile for hours and is roused only for the evening hunting. The male is such a tiny creature that he probably has complete immunity from the attack of the female as he clambers over her grotesque body like a tiny parasite. The need for any courtship in such animated spermatophores should not be very great.

THE MATING

The transfer of sperm is accomplished in a most amazing manner by means of the palpus and the epigynum. Whereas in most animals the copulatory act contributes little or nothing to knowledge of the group, in spiders the details are of great interest and in many cases of deep significance. During the mating the female becomes quiescent and remains in a kind of cataleptic state until its termination. In many species, the female first contributes to the mating by aligning parts of the epigynum so that the corresponding units of the male palpus can be properly oriented. The attitude maintained by the sexes is most constant within the species, and the details sometimes give us data on the general position of the spiders of the series.

Two principal embraces are found among spiders. In the tarantulas and their allies, the six-eyed spiders, and the aerial web spinners, the male usually approaches the female from in front and, moving underneath until his cephalothorax lies beneath her sternum, applies his palpi directly. This is frequently referred to as the *Dysdera* embrace. Among the web spinners it is a quite favorable position since the female hangs inverted below the male and does not greatly menace him. On the other hand, the male wandering spiders that use this type of embrace are in a dangerous situation beneath the jaws of the female. Retreat after the mating, when the female has largely lost her sexual ardor, is likely to be hazardous. In many instances, the males carefully disengage themselves and then leap back and away quickly, showing that they have become conditioned to compensate for a changed attitude in their mates.

The second position, the *Lycosa* embrace, is the one used by the wolf spiders and the running spiders of the higher families. Here the male crawls over the body of the female and, with head pointed in the opposite direction from hers, reaches around the wide of her abdomen to apply a palpus to the epigynum. There is far less danger to the male when he assumes this position, and he is more or less in command of the female until he disengages the palpus and runs away. He serves the right side of her epigynum with his right palpus, and swings around to the other side when using the left palpus. Some of the vagrants have so modified their bodies that it has been necessary to change the type of embrace. Thus, the abdomen of a stocky female crab spider is often so wide that the male must crawl to a ventral position in order to reach the epigynum with his palpi.

The actual copulation, accomplished by means of secondary genital structures, consists in the orientation and pressing of the embolus into the atriobursal orifice of the female. Many primitive spiders that use the *Dysdera* embrace apply both the palpi simultaneously to the orifices just beneath the genital furrow. Since both palpi are applied directly from beneath, it follows that the *right* palpus enters the *left* orifice, and the *left* palpus the *right* orifice, of the female. Other spiders of that series, for example, the tarantulas and their allies, apply each palpus alternately, but probably use the same side for insertion as indicated above. In all the higher spiders this situation is reversed, with corresponding palpi serving the corresponding female orifices. This most interesting fact indicates that specialization in the epigynum and palpus has been accompanied by this profound change in the insertion of the embolus.

The actual union of the secondary genital structures may be of very brief duration, only a few seconds, or it may be prolonged for several hours. When the organs are highly complicated, the insertion is apparently aided by preliminary lubrication of the palpus accomplished by drawing it through the chelicerae, and is finally consummated only after the manipulation of several different elements. The palpus may be scraped across the epigynum until a spur on the tibia, on the tarsus, or on the bulb itself becomes fixed in a particular groove. Once firmly anchored in this starting point, the palpus swings to assume a position that, with the aid of ridges, grooves, and other processes on the epigynum corresponding to its own outline, makes it possible to guide the embolus to exactly the right point for entering the orifice. At this stage, the bulb of the palpus is still largely in its resting position, lying folded in the cup of the tarsus, and the preliminary contacts serve to hold it firmly in place. Most spiders have at the base of the bulb various thin pouches, or hematodochae, that swell up with the influx of blood until they attain enormous size. This distention causes the entire bulb to turn on its axis, which action forces the embolus into the appropriate opening. The whole embolus (usually a thin spine or heavy spur but often a coil that may consist of several loops) is screwed into the epigynum, following the corresponding tubes in this organ until it reaches the seminal receptacle. Semen is then pumped into the chamber by means of strong blood pressure in the palpus, brought about by contractions of the muscles of the body.

The female retains the viable sperms in her receptacles for long periods and dispenses them at the time of egg laying. Although the epigynum has two separate pouches without communication, it is not necessary for the male to supply both of these pouches with sperms to accomplish the impregnation. This usually happens, however, inasmuch as the male exhausts one palpus and then applies the other to the alternate orifice. Those spiders that apply both palpi simultaneously charge both receptacles at the same time.

Polygamy is the rule in spiders, though habits vary. The female and male, if he escapes safely, may pair again. After an initial copulation the female may reject forcibly any male that approaches her, or may submit many times to various males, even after her eggs have been laid. In some spiders it is probable that only a single coition occurs, since the epigynal openings become blocked with a tough plug following mating. This is especially noticeable in the commensal comb-footed spiders, *Argyrodes* and *Rhomphaea*, in which the epigynum is capped with a hard conical cover. In many small species of *Araneus*, notably those of the *mineatus* and *juniperi* groups, the separate epigynal openings are plugged with a black material so tough that in some instances it has been described as an integral part of the epigynum. In the mated *Peucetia viridans*, the green lynx spider, the epigynum is covered with a hard blackish layer, probably composed of dried semen and collateral liquid, and there is usually present a small process from the male palpus, broken off during the mating—a fact which aids greatly in associating the proper male and female when there are more than one species of *Peucetia* in a particular region.

PALPUS AND EPIGYNUM

The pedipalps of spiders are the second pair of appendages of the head and lie on each side of the mouth, being inserted behind the chelicerae. They are six-segmented organs consisting, from base outward, of coxa, trochanter, femur, patella, tibia, and tarsus, and are thus lacking the additional metatarsal segment present in the legs. The basal segment is the coxa, and it usually bears a conspicuous lobe, the endite or maxilla, that lies at the side of the labium and serves as a cutting and crushing instrument during feeding. The remainder of the pedipalp is the leglike palpus, whose tarsus is ordinarily armed in females with a single terminal claw, while in the males it is enlarged and transformed into a copulatory organ.

The spider's palpus (Fig. 3, A, B, F, G) is undoubtedly one of the most unusual intromittent organs that has been developed in any group of animals; its parallel is found only in crustaceans, in dragonflies, and in the arachnids of the order *Ricinulei*. The whole process of its employment is a complicated one, requiring detailed, elaborate acts and routines before it is successful. What do we know about the development of this curious mode of copulation? There is little recapitulation of its origin and evolution in the lives of spiders themselves. Nature usually accomplishes new things by small steps and leaves behind traces of the path that has been followed, omitting missing links to be filled in by speculation. Various araneologists have attempted to outline the various events leading to the spider palpus and a plausible path is the one outlined below.

The aquatic precursors of the arachnids probably fertilized their eggs by spraying them with semen while they were being deposited by the females, perhaps in the fashion of our horseshoe crabs. When these emergent arachnids left the water for an aerial environment, a new method had to be devised to protect the sex products from desiccation and safeguard them during transfer to the female. Semen merely dropped on the substratum in the old way could have been lost. The males learned to produce the sperm globule at the right time and to convey it into the uterus for quick fertilization of the eggs or storage for later use. The problem was solved in various ways by different arachnid groups. The male solpugid, one of the most generalized of all arachnids, seizes the female, and by pinching her abdomen, causes her to fall into a state of torpor. Then he ejaculates a sperm globule and forces it into her gonopore by means of his chelicerae, an effective and direct way of handling the problem. In some other arachnids various kinds of secondary vehicles, called spermatophores, have been developed

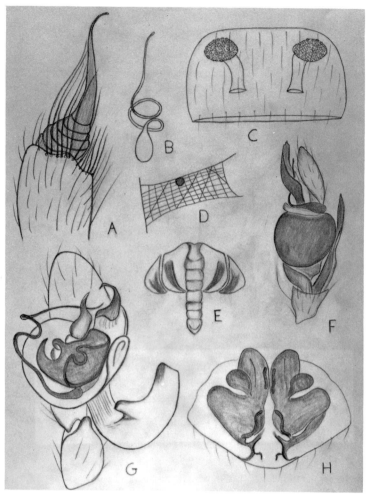

Fig. 3. The male and female genitalia. A. Palpus of *Filistata tractans*. B. Outline of seminal receptacle of typical male palpus. C. Epigynum of *Filistata tractans*. D. Sperm web of male spider. E. Epigynum of *Araneus marmoreus*. F. Male palpus of *Pachygnatha autumnalis*. G. Male palpus of *Nesticus hoffmanni*. H. Epigynum of *Nesticus hoffmanni*, internal view of seminal receptacles.

to cradle the sperm globule until it can be passed on to the female. The males of scorpions, pseudoscorpions, and whip scorpions of various orders extrude spermatophores from the genital orifice and attach them to the substratum. The spermatophore is often a slender or thick stalk at the tip of which is placed the sperm globule. In scorpions the tip of the spermatophore is provided with an injecting mechanism that forces the sperm into the female gonopore. By various stratagems, described under the term courtship, the males maneuver their mates into favorable positions where the sperm globule could be taken into the gonopore by the females.

It now seems likely that spiders first voided a sperm globule and then transferred it directly to the gonopore of the female by means of the leglike pedipalpi. The next step was modification of the claws at the end of the pedipalpi into organs for storing and injecting the semen. It seems unlikely that sperm induction into the palpi from a droplet on a formal sperm web came before successful development of the storage vessels. In other words, the sperm web (Fig. 3, D) is likely not a remnant of a fading spermatophore but a practical development necessary when spiders became confirmed to life on webs. Having the semen secure in a reservoir does away with the need of ejaculating it during or immediately preceding the mating, and lessens the risk of losing the female during such preliminary routines. The intermediate stages of the araneid mode of copulation have been dropped out and do not now persist in the memory of the race.

The palpus of the male spider (Fig. 3, A, F, G) only gradually developed into the complicated organ we now observe. At first it resembled that of the female, and was armed at the end with a single tarsal claw that picked up the sperm globule and pressed it into the female gonopore. Gradually the claw became transformed into a cup-shaped receptacle, from which the liquid was less easily lost, and finally the cup was closed at the end until only a small opening remained through which to take in and drain out the semen. This receptacle (Fig. 3, B) is the all-important element of the palpus, and in its simplest form is made of three more or less well-defined parts: a basal expanded portion termed the fundus, a coiled intermediate tube called the reservoir, and the delicate terminal ejaculatory duct. We see this elementary receptacle in the palpi of many spiders where it has remained very simple, and can still discern it as the prime element of those spiders in which the structure has become greatly elaborated.

At first, the receptacle was appended to the tarsus as a simple extension, but in this position it was quite liable to be broken or injured in some way. Specialization has proceeded to protect it and its delicate terminal duct, and to make it more effective as an intromittent organ. Around it has been developed a protective cover called the bulb. The tarsus itself has been excavated to form a protective receptacle in which the whole organ can lie when at rest. Muscles and blood pouches have been evolved to make possible the ejection of the semen. On the bulb itself have arisen processes that are used to orient the parts of the palpus in relation to those of the epigynum, and apophyses on the tibial and other palpal segments act in a similar manner.

As is true for the male, the gonads of the female are hidden deep in the abdomen. From the two ovaries come off oviducts that join to form a single tube, the uterus, which opens externally through a transverse slit at the middle of the epigastric furrow near the base of the abdomen. It is probable that in ancient spiders the palpal organs of the male were pressed into the wall fronting the upper part of the uterus and gradually imprinted themselves as primary sperm-holding vessels. These invaginations gradually were formed into the pair of sclerotized spermathecae for receiving the semen and later returning it to the uterus at the time of egg-laying. These spermathecae are the homologues from which develop the external epigynum. The tarantulas and their allies and many of the primitive true spiders have relatively simple male and female genitalia and in these the epigyna are of this simple type, often termed *haplogyne*. During the mating, the male introduces his palpi, frequently simultaneously, into the openings hidden under the lip of the genital groove (Fig. 3, C). Only the terminal part of each palpal organ, the embolus, is pressed into the spermathecae. In higher spiders the primary fertilization tube of each internal spermatheca is divided into two, one part remaining as the internal fertilization canal and the other part migrating to the surface to form an external opening to receive the embolus of the male palpus; this is termed the *entelegyne* condition (Fig. 3, D, H).

The epigynum, whether internal or external, is composed of two essentially symmetrical independent units, each of which serves as a sheath for the male embolus of its particular side. There is a close correspondence between the physical proportions of the duct leading into the spermatheca, and the embolus of the palpus, inasmuch as they have grown up together. In the higher spiders entrance to the spermathecae is gained through a pair of external openings. Much of the perfection and elaboration of the palpi and epigyna must be attributed to this new external position of the orifices, which makes possible the adoption of a different mode of pairing, in which the male is less vulnerable.

The external epigynum (Fig. 3, D, H) has become specialized in various ways. In many spiders there is a convex atrium surrounded by a distinct rim within which lie the orifices. These latter are separated by median ridges that guide the embolus into its particular channel. At the front or behind may be a hood or tubercle articulating with an apophysis of the palpus. A conspicuous finger often overhangs the atrium and serves to fix the palpus in just the right position to make the pairing possible. Some epigyna are remarkably complex but it should be noted that not every apophysis of the male palpus has been developed to fit a corresponding depression in the female epigynum. As in all structures of animals, the tendency to become more elaborate may go far beyond the needs of the animal. Many of the spurs and strange projections in complicated palpi may prove only useless luxuries that contribute nothing to mating.

The very close correspondence between the male and female genitalia of insects led Leon Dufour to the formulation of the so-called "lock and key" principle. It was his belief that the crossing of species was impossible for physical reasons, and that the male organ could not be introduced into that of a strange female because of differences in length, shape, and size. The female organ was regarded as an unyielding lock that could be opened only by a key that corresponded exactly with its form. Whereas we must reject the theory that these organs are adaptations that exclude the crossing of species, and instead assign that function to fundamental instinctive patterns probably based on chemotactic stimuli, it must be admitted that in spiders the differences between the genitalia of remote or even allied groups are usually sufficiently great to make pairing impossible—in effect a "lock and key" presenting an impassable barrier to all but the most closely related species.

It must be kept in mind that the secondary genitalia of spiders are extremely ancient organs probably fully evolved long before the late Paleozoic era, where we find fossil spiders. Both primitive true spiders and living tarantulas, discretely separated even at that time, have similar palpi, indicating that the general features of their organs antedate the separation of the two suborders. It is little wonder then that in the palpi and epigyna are clues to the general phylogeny of the whole group. These organs have undergone changes corresponding closely with the specialization of spiders themselves. Indeed, sexuality and the araneid mode of copulation are adaptations that have probably contributed more to spider evolution than have any other features.

6

The Evolution of Spiders

Spiders owe much to the past. The proud jumping spider of today, attired in flowing robes of ermine and crimson and with great smoky eyes intently following every movement of a gleaming bluebottle fly, bears little resemblance to its reserved, myopic forebears. The sedate orb weaver, hanging from a web of wondrous design, has come a long way from the clumsy land creature that first attempted to climb into the shrubs. So changed are many spiders that we can scarcely discern in their bodies any clues pointing to their origin.

From fossil evidence we know that spiders are ancient creatures, and that they were confirmed land animals long before the vertebrates had become free of the bondage of aquatic life. A large part of their evolution must have been undergone during the Devonian period, which has left one record of an enigmatic or perhaps spurious spider, *Paleocteniza*, from the Rhynie Chert of Scotland, occurring with mites and numerous excellently preserved arachnids of the extinct order *Anthracomarti*. Splendid fossils come from the coal measures of the Carboniferous era, in both Europe and the United States, revealing that at that time highly developed, typical spiders were already in existence. Much remains to be learned of these earlier araneids, and of the arachnid group that gave rise to them, since we have no evidence to show that spiders have been derived from any other living or extinct group of arachnids. Nor do we have any conclusive evidence that the arachnids evolved from any particular group of arthropods. The classical theory of Ray Lancaster, which postulates the trilobites as the ancient group from which have been derived scorpions and typical arachnids on one hand, and eurypterids and king crabs on the other, has long been severely criticized. More recent evidence, however, strengthens this general thesis and points to the derivation of these diverse arachnid groups from relatives of the conservative trilobites living in the Cambrian seas. An alternate theory would have the arachnids derived from some land creature, similar perhaps to the sluglike *Peripatus*, but there is little evidence for this at the present time.

The phylogeny of spiders has long been the subject of much speculation, and there is still no general agreement as to the fundamental paths that were followed. This volume attempts to lay down only the broad features of spider evolution, and acknowledges the inclusion of much speculative matter. The phylogeny of any group can be postulated by means of the fossil record, and also by aids from taxonomic classifications, which are frequently indicative of the racial history of a group.

The ancestral stock from which come all major spider groups originated some time before the Carboniferous era. These creatures probably bore a close resemblance to the spiders fossilized in the coal measures, with abdomens enclosed in hardened plates—wide tergites above, sternites below, and hard, narrow pieces (pleurites) on the sides. Four pairs of similar fingerlike appendages were present beneath the abdomen at about the middle. Just in front of these appendages was a pair of spiracular openings leading to book lungs, and a second similar pair was present farther forward, at the base of the abdomen. Inside the abdomen was an elongate heart into which opened five, six, seven, or even more pairs of ostia, each pair representing a segment and those at the rear much reduced in size. Thereafter, the tendency was to increase the size of the organs in the anterior segments and gradually to suppress the posterior ones, resulting in the loss of some of the ostia and the supporting muscular systems. The gradual reduction and loss of the units of internal segmentation were matched in the external plates, which resulted in an actual migration forward toward the spinnerets of the anal tubercle.

The cephalothorax of the ancestral spider was relatively long as compared with its width, and was marked by a longitudinal median groove. At the front end were eight eyes set close together on a low tubercle. The legs were of moderate length, quite stout, and each tarsus had at its tip three tarsal claws, the outer paired ones relatively long and smooth, the inner unpaired one short and only slightly curved. The chelicerae were large, set parallel to the long axis of the body, with robust fangs. The gland in the basal segment secreted a weak venom, largely unnecessary since to subdue prey reliance was placed mainly on the strong legs and sharp fangs.

The earliest spiders were cautious hunters that groped about on the ground and made little effort to establish a permanent station of refuge. Food perception was accomplished by sensory leg hairs which tested the terrain, inasmuch as their small eyes were useful mainly to distinguish light from darkness. These sluggish prototypes lived a timeless life of leisure on the tangled jungle floor of their humid swampland. Only during molting and egg laying was it desirable to be concealed from the few wandering predators, and from less worthy adversaries that under those trying circumstances might do injury to the eggs or to the spiders themselves. The first step toward a life of dependence on silk was the coating of the eggs with excretory material from the abdomen, voided by coxal glands that opened through the abdominal appendages. As the product of the glands became more suitable for use as a gluing and covering agent, and the spinnerets more adept in their application of the gummy liquid, greater possibilities for the use of the crude silk opened on all sides.

These ancient spiders were perennials. Each female produced and cared for many egg masses during her life with varying degrees of efficiency and success. Those survived that were more adequately protected by a silken cover, and guarded in long vigils by the mothers, whose regard for the safety of the egg mass was being tried and modified by an increasingly hostile and enterprising band of predators.

These ancient spiders persisted in the Paleozoic and in the coal measures we find them little changed from their precursors. During that time when much of North America was a dismal, swampy area covered by forests of strange plants and trees, there lived in the region of modern Illinois similar spiders whose abdomens were armored with hardened plates. During the same era such spiders were also found in Europe, more numerous in species and so much more diversified that the imprints seem to belong to several distinct types. All have well-marked tergites. The Illinois spiders from the Pennsylvanian shales of Mazon Creek are placed in the genus *Arthrolycosa* of the family *Arthrolycosidae*, and they are remarkably like the modern liphistiids. However, nothing is known of their spinnerets, claws, sternum, or of other features largely used in classification. It is now believed that the only surviving representatives of such ancient spiders are the small family *Liphistiidae*, found now only in eastern Asia. They and their forebears were given the subordinal name *Mesothelae* by Pocock in 1892 and this has reference to the presence of eight spinnerets in median position on the abdomen.

All other modern spiders are now believed to be derived from the *Mesothelae* and were placed in a single group *Opisthothelae*, also by Pocock in 1892, and in these the spinnerets are situated farther behind near the anal tubercle. By the late Paleozoic era the two principal groups of spiders known today had been developed: the *Mygalomorphae*, or tarantulas and their allies; and the *Araneomorphae*, the true spiders. Discretely separated already during the coal measures, these two lines have grown up together and in many ways their accomplishments have paralleled each other, a natural development since both were originally endowed with similar equipment and potentialities. However, for various reasons the true spiders surpassed the tarantulas during the Tertiary period and became the dominant group.

From creatures like *Arthrolycosa* and its European cousins have come all the modern spiders known as trap-door spiders, purse-web spiders, funnel-web tarantulas, and typical tarantulas in their great variety; in short, all of the *Mygalomorphae*. The tarantulas were present in the Paleozoic era with plates on the back of their abdomens and many of the tarantulas have retained well-marked evidences of dorsal segmentation through three or four hundred million years until the present time.

Certain shadowy forms from the Carboniferous era, contemporaneous with the older tarantulas, have been assigned with some confidence to the *Araneomorphae*, or true spiders. They appear to lack or have few hard plates on the abdomen, and to assume—in a vague way at least—the form of some of the higher spiders. In what ways do these emergent creatures, from which is derived the vast array of modern true spiders, differ from the Paleozoic tarantulas? How did the branches separate?

The fundamental change may well have been one of behavior, a change in habit or attitude rather than a physical alteration. In some way it is related to their greater use of silk, to their more expert spinning, and to the retention of the anterior median spinnerets as functional organs until the principal lines of true spiders were well established. Associated with this divergence from the tarantulas was the gradual change in position of the chelicerae. In modern representatives the chelicerae are now twisted at right angles from the long axis, called diaxial because the fangs now point toward each other. This freed the spider from throwing back his carapace to put into use chelicerae with fangs pointed forward in paraxial position. Improving eyesight and the new position of the chelicerae made the capturing of prey more effective. Cutting edges were being developed on the coxae of the palpi to aid in crushing the body of the prey. The venom was becoming more potent and the voluminous glands were pressing beyond the limits of the cheliceral segments into the head itself.

Loss of the heavy abdominal plates was another consequence of the change in life. This armor disappeared gradually and still is vaguely indicated in a very few modern true spiders. *Paleodictyna* from the Baltic amber retains conspicuous plates and this tardy divestiture suggests the possibility of finding many more fossil true spiders retaining dorsal plates.

The course of true spider evolution has been charted largely by silk spinning. The *Araneomorphae* began their history with the same equipment as the sister tarantula group—eight functional spinnerets of nearly equal size. But whereas the tarantulas were content to spin in a modest way, the true spiders began to use silk more often and with greater efficiency. Since the lateral spinnerets undoubtedly were bisegmented at an early date, and had the advantage of greater length and strategic position, it was natural that these should be developed and improved at the expense of the unisegmented median pairs. The great reduction in size and early loss of both anterior median and lateral spinnerets in all but a few relict mygalomorph spiders reflect their failure as spinners. The true spiders, on the other hand, retained all these spinnerets for a long period, and some still keep the anterior median pair. In this connection it is worth noting that the metatarsal comb—the calamistrum—used to brush across the spinning field of the median spinnerets, was in all likelihood an early invention, and that all true spiders once spun cribellate threads. The retention of the anterior lateral spinnerets as prime spinning organs, probably made possible by persistent use of the incipient cribellum, was the key to true spider superiority, and actually caused the divergence of the true spiders from the parent line.

It was inevitable that, in addition to the formal silken covering over the egg mass, many threads would be scattered more or less haphazardly from this spinning center. Such wild lines were instrumental in giving to the mother spider another advantage in her efforts to guard the eggs, communicating the approach of an interloper by vibrations on the threads. The range of touch perception was thus in one step expanded far beyond mere contact with the sensory hairs on legs and body; the deadly predator or the blundering insect often became the prey of the vigilant spider. In this two-dimensional maze of threads, with the egg sac as the central theme, was the germ of all the webs that have made the true spider dominant.

The stringing out of silken lines continued during the whole life of the spider, as well as at the egg laying, and has continued to the present time as the dragline habit of modern spiders. With a secure line attached to the spinnerets, the spider could now venture upon precipitous surfaces with a certainty of quick recovery from falls. Since the dragline of true spiders is ordinarily spun through the anterior lateral pair, the tarantulas, in suppressing these spinnerets, virtually precluded the future possibility of becoming aerial spiders.

The lifeline of the whole group of true spiders became their silken threads, and those that refused to accept subservience to this material died out. Every spider became sedentary to a degree and none has been able to divest itself completely from silk since those early days. Each major group of spiders diverged from the others with essentially the same type of spinning equipment, and with a well-founded instinctive knowledge of silk spinning. In each of the lines similar types of webs and traps for the capture of insects have been evolved separately.

In one group the anterior median spinnerets have been perpetuated in a modified form as the cribellum. These creatures come down to modern times in a more or less homogeneous line as the cribellate spiders. The whole series probably diverged quite early from the main stem and, although their physical features mark them as a more generalized group, they have done remarkable things with their heritage. All the remaining true spiders lost the anterior median spinnerets, but in most of them vestigial evidences can still be observed.

During the early history of the ecribellate true spiders, a trend—already running a similar course among the cribellate types—toward the simplification of the various organs was operative. The mutations began at different times and progressed at different rates, so that in modern types generalized features sometimes exist side by side with profound specializations. The tendency has been to simplify the fundamental systems, to make fewer segments and functional units (such as book lungs, tracheae, ostia, spinnerets) do the work of the greater ancestral number.

The abdomen was developing into a highly developed center for silk spinning and in most lines tended to become shorter; in some, globose. The spinnerets gradually attained a position at the tip of the abdomen, near the anal tubercle itself, indicating the virtual reduction of the abdominal segments to four. Some of the spinnerets later became elongated and modified to perform special types of weaving, and others became so reduced in size that in certain cases only the anterior lateral pair remain as functional spinning organs. A notable achievement was the transformation of the hind pair of book lungs into a pair of tracheal tubes soon after separation from the cribellate line; this development was followed in most spiders by fusion of the openings into a single tracheal spiracle. In a few lines the front pair of book lungs was also converted into tracheae. The number of ostia in the heart was reduced from four to three pairs, in some species even to two pairs, and the remaining ostia assumed the function of the lost members.

Changes of many kinds were also taking place in the cephalothorax and its appendages. Especially notable was the migration of the eyes from the original local center at the front edge of the carapace, to the sides and to other positions of greater advantage. The anterior median pair was lost early in a whole group of spiders that persists until the present time as six-eyed spiders, and whose other characteristics indicate that they are among the most generalized ecribellate true spiders. Other types enlarged their eyes, and with appropriate changes to the legs and body, came to place considerable reliance on sight as an aid in gaining a livelihood.

The early ecribellate spiders were at first terrestrial types that stalked over the soil and low vegetation in an upright position, trailing their dragline threads behind. Some of the lighter ones began essaying trips into the herbs and shrubs, and learned to use their claws to climb from twig to twig, hanging back-downward from their lines. The third dimension was becoming a spacious reality to these extroverts, and with its spaciousness came freedom from attack by ground creatures. The egg sac was installed in the center of the tangle of threads, completely safe from flying predators, which could not reach it without becoming enmeshed in the lines. And from entangled insects of many kinds the spider was securing its food. The aerial web spinners became specialists for a life on silken lines, modifying the unpaired claws of the tarsi into effective hooks to cling to their threads.

Many spiders remained creatures of the soil, and for running or climbing made little or no use of the unpaired claw. Some of these hunters lost the unpaired claw, developing instead adhesive claw tufts that allowed them greater ease of climbing.

7
The Mygalomorph Spiders

The names of trap-door spider, purse-web spider, tarantula, and sheet-web tarantula bring to mind some of the most famous of all spiders—spiders that rival in size the largest land invertebrates, spiders that have become renowned for their wonderful burrows and handiwork. All are four-lunged spiders belonging to the suborder *Mygalomorphae;* they are often referred to as mygales but in this book are collectively known as "tarantulas" or mygalomorph spiders in contrast to the "true spiders" of the suborder *Araneomorphae.*

The mygalomorph spiders are more generalized than the true spiders and are ancestral to them. As a group they are longevous, all living more than a single year and some of them attaining great age—as age is measured in invertebrates: up to or even exceeding twenty-five years. They are large, probably averaging more than an inch in length as compared with less than one-fourth that size for true spiders. Some of the typical tarantulas attain a body length of three and one-half inches; at the other end of the scale, the pygmies, the tunnel and sheet weavers of the genus *Microhexura*, are one-eighth inch long. Along with great size the mygalomorphs perhaps retain as a consequence the second pair of book lungs and other features correlated with their primitive station among spiders as a whole.

Although it must be conceded that the true spiders have attained a higher degree of development—as evidenced by their greater numbers, variety of structure, and multiplicity of habit—the tarantulas should not be thought of as vastly inferior. They have become notably specialized in their own way, and in instinctive behavior have nearly kept pace with their cousins.

The most important single character that distinguishes the mygalomorph spiders is the articulation of their chelicerae—termed paraxial as contrasted with the diaxial position of true spiders—and other details of their mouth parts. The chelicerae (Fig. 5, B; Plate VIII) are robust and two-segmented, as usual, but with their long axis parallel to that of the body, and with movement in a vertical plane. As befits these powerful spiders, the fang of the chelicera is a stout, curved weapon. In order to drive the fang into the victim, the body must be elevated. These creatures strike with great speed, but because of their poor eyesight and the necessity for wasted motion, their method is probably inferior to that of the true spiders. When confronted by man or any creature outside its normal experience, the tarantula throws itself back with unsheathed fangs and maintains itself in a position of readiness to strike. This is a defensive attitude, but also one favorable for attack.

The venom glands of the mygalomorph are entirely contained within the basal segment of the chelicera. Since its offensive needs are met by a powerful body and robust jaws, the necessity for great quantities of potent venom is minimized. In most tarantulas the coxa of the palpi lacks the endite or maxilla, an expansive lobe used in crushing and cutting the prey.

All the *Mygalomorphae* have two pairs of book lungs, clearly visible on the ventral surface of the abdomen and notable for their large size. Only one family of true spiders, the *Hypochilidae*, has retained this four-lunged condition, and they are the most generalized of all true spiders in many other respects as well.

A moderate number of mygalomorph spiders range up into the temperate zones, but the group is essentially tropical and subtropical in distribution, about 1,500 species being known from these zones all around the world. During the Paleozoic era, their ancestors dwelt in the swampy, humid forests that became the coal measures of the United States and Europe. No tarantulas are known from the Mesozoic era, but we can be sure that they were well represented, and perhaps at that time equaled the true spiders in numbers and variety. Because of their secretive habits, which have resulted in a meager fossil record, few Cenozoic mygalomorphs are known; small numbers have been found in the Baltic amber of the Oligocene epoch, and in the Oligocene shales of the Florissant formation in Colorado.

At some time during the early history of the *Mygalomorphae* the line split into two principal branches, which have descended to us side by side as our modern fauna. On the one hand are the typical tarantulas and the trap-door spiders; they represent the largest and best known series. The second group is somewhat inferior in physical equipment (if we measure this in terms of the degree of change from ancestors), and has come down as a reminder of what most mygalomorphs probably were like during the Mesozoic era. These latter we refer to as the atypical tarantulas.

THE TYPICAL TARANTULAS

In this series, which includes the tarantulas, the sheet-web tarantulas, and the true trap-door spiders, there is no visual evidence of dorsal segmentation of the abdomen. The endites are not at all developed in the American species, but in some exotic forms a small angled spur or a well-developed process may be present. Nearly all have only four spinnerets, the hind lateral and median pairs; these are located close in front of the anal tubercle. The commonly associated characteristics of tarantulas–large size and hairy covering–should not mislead one in identifying members of this group. Many are relatively small in stature. Only the wandering hunters, the true tarantulas, are thickly clothed with velvety wool and long silken hairs; others appear quite naked by comparison.

A few of the mygalomorphs have become vagrant, but none has attained the degree of freedom enjoyed by certain true spiders. Failure to improve vision has resulted in the development of few accomplished runners, jumpers, or climbers, and none of these tarantulas has become dependent on silk as have the aerial true spiders. Their reliance on touch is perhaps even stronger than in the araneomorphs; the hairy covering of the vagrants, for example, serves admirably to make them aware of the presence of prey.

The typical tarantulas have been most successful in living a secretive life hidden in the ground, with the consequence that many have become specialists in subterranean existence. Their general makeup fits them eminently for a successful life in tropical regions, where competition may not be so keen. Few Americans realize that in the southern and western portion of the United States exists a rich and varied fauna of mygalomorph spiders, eighty or ninety species including many with curious habits.

Trap-Door Spiders. Many spiders tunnel into the soil, but the true trap-door spiders of the family *Ctenizidae* are the most accomplished burrowers and the most gifted artisans. They and their relatives can claim to be the inventors of that superb device to ensure privacy, the trap-door, for they represent a stock that was probably capping burrows with doors long before many true spider emulators were evolved. The first description of this interesting device was given by Patrick Browne, who in 1756 illustrated the nest of a West Indian species in his *Civil and Natural History of Jamaica.* A few years later the nests of *Nemesia* were described from France, being likened to "little rabbit burrows lined with silk and closed with a tightly fitting movable door." Although trap-door spider nests continued to attract popular attention thereafter, it was not until 1873, when J. Traherne Moggridge published his careful studies on the habits of these animals, that any comprehensive treatment was accorded them.

Moggridge was able to distinguish four distinct types of nests among the species he studied. The first nest was a simple tube, a cylinder closed with a thick, beveled door, which he termed the "cork door"; the second nest was a simple tube closed with a thin or "wafer" door; and the third type of nest was a simple tube with a thin outer door and a second door part way down. Moggridge's fourth classification was the most complicated: a nest capped on the outside by a thin door, and having an oblique side tunnel, connected with the main tube, and at the entrance to which was a trap door. Several other types of nests have since been discovered in various parts of the world, some of them much more complicated than those described by Moggridge. Furthermore, the distinction between the cork door and the wafer door, while valid enough in the extremes of each type, gradually disappears as we examine long series of intergraded nests.

The true trap-door spiders have developed a comblike rake of large spines on the margins of their chelicerae, and this they employ as a digging instrument. With its aid they are able to cut and scrape small particles of earth, which they then mold into balls and carry outside the burrow. They waterproof the walls of the tube by applying a coating of saliva and earth, so that the surface becomes smooth and firm. Then they apply a silken lining of variable thickness and extent, in some cases not fully covering the burrow, while in others coating the whole tunnel with a thick fabric.

When the maturing spider outgrows its burrow, it enlarges it by cutting and scraping off bits of earth with its rake and carrying them away from the site. Rocks embedded in the soil may oblige the spider to pursue a tortuous course, or to dig a new tunnel in a more favorable situation. The spider rarely deserts its burrow voluntarily. When forcibly removed, the spider will accept the unoccupied tunnel of another spider, or a cavity especially made for it, and then proceeds to remodel this in its characteristic way to suit the pattern of previous homes.

Although spiders of many other families burrow, the trap-door mygalomorphs have far outstripped them in the excellence of their tunneling. They have become specialists that dig with better instruments, line with greater care, and are the originators of the practice of capping the burrow with a perfect lid. While the trap door is not a unique accomplishment of these spiders, having been developed independently by several other groups, it bespeaks a mastery not closely approached by an emulator.

The typical burrow is spacious enough in part of its length to allow the spider to reverse position at will. Within its confines the spider finds a haven until violent or natural death. What are the advantages of this abode, which has become such a dominant element in the lives of these spiders? In the first place, it is the property of a single, unsocial individual and can become, with the passage of time, more and more adequately coated with silk, more and more familiar in its every part, and thus increasingly acceptable to the spider. It is a retreat from the rays of the sun, the extreme heat of which is shunned by nocturnal and diurnal forms alike. Its hinged lid, which can be opened or closed at will, prevents rain and surface water from entering, thus keeping the nest drier than surface situations. Since all the burrowing spiders live more than a single year, the tunnel serves to temper the extremes of inclement weather over long periods. The tube beneath the surface is cooler during the summer heat, and somewhat warmer during the winter cold. Relatively inconspicuous because of its location flush with the surface of the ground, the burrow opening may be made more difficult to discern through the efforts of the spider. During the hottest part of the summer, when raiding parasitic wasps are present in maximum number, the opening may be closed tightly with earth and silk. Mosses, leaves, sticks, and other debris can be placed to advantage on the lid and around the entrance, the result—to human eyes at least—hinting of camouflage. When in active use, the burrow can serve as an ambush from which the spider rushes out to seize its prey; and once an insect is caught, the nest becomes in most cases the dining room. At the proper season the burrow may also serve as a mating chamber, and a nursery when the eggs are laid and enclosed in their sac hung up on a wall. Later it becomes the home of the young spiderlings, often for many weeks after their emergence from the egg sac.

The opening to the surface is the spider's only contact with the outside. It is the vulnerable element in the circumscribed abode, but at the same time it allows the occupant to be menaced from only one direction. On the surface, an inferior sensory equipment places the trap-door spider at great disadvantage in combat with its specialized enemies. Within the burrow, it faces the enemy protected by a silk door, and should that be torn away, it still has a favorable situation for the use of its strong fangs.

While demands for privacy have probably inspired the perfection of the underground castle of the trap-door spider, it is intriguing to think of the domicile in terms of response to the ravages of some arch enemy. By far the most fearsome assailant is the spider wasp, a common name for various species of *Pompilidae*, which are exclusively spider predators. Other enemies may wreak their toll in an insidious way and possibly destroy more individuals than does the wasp, but this gleaming tyrant is a predator of the first magnitude whose prey is the large, adult spider and whose victory is won in hand-to-hand struggle.

Actively foraging over the soil, unerringly directed by a sense not conditioned by previous experience, the wasp arrives at the trap door, beneath which sits the prospective victim—possibly aware by then through its delicate tactile sense, of the presence of an intruder. If unprepared, or if its resistance is finally broken down, the spider quickly finds itself confronted by an enemy that has lifted the trap door or gnawed through it and entered the burrow. The struggle that ensues is not a battle of giants. It is a very unequal one from which the wasp almost always emerges the victor. Swift and sure in movement, liberally endowed with fine sensory equipment, and armed with a deadly sting, the wasp confidently assails a larger creature fighting on a prepared battleground in the deep recesses of its burrow. After a brief struggle the wasp paralyzes the spider with venom from its sting, and then proceeds to deposit on the spider's abdomen an egg, from which will hatch a voracious larva. Doomed to lie helpless while furnishing fresh food for the larva, virtually dead if not actually so, the once mighty spider finds its castle converted into a crypt. Industrial skill has failed to make the burrow impregnable to its most formidable enemy.

During the growing period, when the spider is remodeling and strengthening its closed tube, it is less subject to the attacks of marauding wasps that, in order to satisfy the food requirements of their offspring, pass up the smaller burrows in favor of mature or nearly mature prey.

We pass now to consideration of the three better known types of trap-door spiders found in the United States. The first of these spiders constructs the classical type of nest that Moggridge called the "cork nest." The most notable nest of this type is made by *Bothriocyrtum californicum*, the common trap-door spider of southern California (Plates II, III, IV). Examples of this nest (Plate III) are to be found in many collections, and may be purchased from various biological supply houses. It is the typical nest with a thick cork door and is illustrated in many works on natural history. Another group of spiders that is widely distributed across the United States from the East Coast to Arizona, the genus *Ummidia* (formerly *Pachylomerus*) also makes this type of nest. The several species are very handsome animals (Plate 1), with shining black cephalothorax and legs and a dusky abdomen.

The cork nest (Fig. 4, A) is a simple tube without side branches, lined completely with silk. Ordinarily the burrows are shallow, from five to eight inches in depth, with a diameter essentially the same throughout and large enough, especially near the entrance, to permit the spider to turn around. The distinctive feature of this nest is the door. It is made of layers of earth and silk, and is so constructed that it fits perfectly and tightly closes the mouth of the tube "much as a cork closes the neck of a bottle." The cork door cannot stand open; it falls and closes of its own weight, and the tube mouth is beveled to receive it.

In western Florida *Ummidia audouini* digs its burrows in the sides of steep, stream-cut banks in moist and shady ravines. Arizonan and Mexican *Ummidia* favor open spaces in the sun-baked creosote-bush deserts. In southern California *Bothriocyrtum californicum* makes its tunnels on sunny hillsides that in early summer bear a thick covering of native grasses. The spiders that build cork nests are plump animals with rather short legs and a broad carapace. They have rows of short digging spines on the front legs which aid in scraping and cutting the soil. Their rounded bodies fit the burrow snugly, with the legs pressing closely against the sides. Their structure bespeaks strength and ruggedness. As is true of most spiders, they are active during the evening and at night, but they rarely leave their burrows. At the exit of its tube, holding the door ajar, sits the spider (Plate IV) ever watchful for the approach of food. On occasion it will rush forth

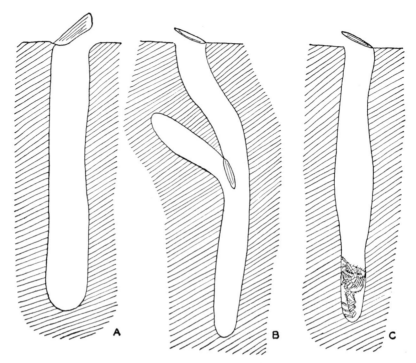

Fig. 4. Trap-door spider burrows (diagrammatic). A. Cork-door nest of *Ummidia*. B. Double-door of *Myrmekiaphila*. C. Nest of *Cyclocosmia* with spider in narrow recess.

to capture an insect, but most of its prey is captured without completely leaving the burrow. Inside the nest it is an agile creature, outside a clumsy one. When disturbed, it closes the door firmly and holds the lid with chelicerae and claws, bracing its legs against the sides of the silken burrow. The species of *Ummidia* and our single species of *Hebestatis* from California have a smooth saddlelike emargination on the third leg, often flanked by stiff spinules, that presses against the lip or side of the burrow. When in this defensive position, considerable force is necessary to dislodge it. Even with the aid of a knife blade one has difficulty in forcing the door.

Three well-known genera, *Actinoxia* and *Aptostichus* (Plate V) of the Southwest, and *Myrmekiaphila* (Plate VI) of the southeastern United States, make the type of nest Moggridge called the "double-door branched nest." The trap door of this nest is of the wafer variety. It is a thin, suborbicular cover almost wholly made up of silk, without layers of earth, and lies on the burrow entrance rather than fitting into the aperture. It is not substantial enough to serve as much of a barrier to an intruder, being soft and pliable, and not heavy enough to fall over the opening if it is pushed back very far. It is only a superficial, hinged cover which may be camouflaged outside with moss, earth, or debris. The burrow proper is a cylinder lined with silk; but the particular innovation in this nest is the second burrow, a secret side chamber cleverly concealed by a trap door so constructed that it can close either the main tunnel or the side branch.

The burrows of *Myrmekiaphila torreya* are found on the leafmold-covered slopes of Torreya Ravine in Liberty County, Florida. This species digs a burrow that averages about ten inches deep. The nests are usually found in sandy soil penetrated by a maze of roots, and almost always contain at least one or more abrupt bends. Halfway down the tubes are the side chambers, one to a burrow, marked by water-type doors. The entrances to the outer burrows are lined with silk and provided with a peculiar type of door which, when standing open, is more like a silken collar than a trap door, but which takes on the appearance of a well-camouflaged door of the wafer-type when closed by a slight push. This door the spider sometimes leaves standing open during both night and day.

Most of the trap-door spiders that make an inner door (Fig. 4, B) are about two-thirds of an inch in length. Their bodies are slimmer than those of *Ummidia* and *Bothriocyrtum*, are ordinarily yellowish brown, and sparsely clothed with brown hairs. Their legs are longer and lack the rows of stout spines that the other group has, possessing instead a light-to-heavy scopula of hairs on the distal segments. These spiders are better able to wander about a bit and move short distances from their tunnels to capture insects.

The open door of *torreya* appeals to us as virtually being an invitation to enter. Atkinson, who studied a similar species in North Carolina, thought that the principal chamber was intended as a prison for ants that wandered in and were captured after closing the inner door. He called the genus *Myrmekiaphila* because his spiders built their nests near or even in anthills, and he believed that the ants made up a large part of the spider's food supply. Moggridge interpreted similar nests in terms of a protective device. A wasp, intent on paralyzing the spider and placing an egg on its body, finds a trap door, which may be open or which she may open by force, and enters in search of the spider. The spider meanwhile rushes to the bottom of the burrow and closes the main tube with the inner trap door. Should the wasp persist, the spider crawls into the side chamber, moving the dual-purpose door to protect the opening. Once the main tube has been fully explored and found empty, the wasp may leave without discovering the inner chamber.

Cyclocosmia is a genus of trap-door spiders remarkable for the strange shape of its abdomen, and notable for its discontinuous distribution. Two allied species occur in our southeastern states, a third in eastern Mexico, and a single giant species in Indochina. Thus, our *Cyclocosmia* enjoy the distinction, along with the American alligator and various other animals and plants, of having their nearest relatives in Asia.

Cyclocosmia truncata (Plate 2) is a large, fat creature rather closely related to *Ummidia* except for the abdominal structure; it also lacks the polished, gripping saddle on the third legs of the latter genus. The round, leathery, caudally truncated abdomen of *truncata*, in the absence of actual observations, led to intriguing conjectures as to what use it was put by the spider.

The initial description of *truncata* was made by Nicholas Marcellus Hentz, the father of American araneology, who in 1841 gave it the name of *Mygale truncata*. His specimens, all of which were females and all since lost, came from Alabama. In his words: ". . . this spider dwells, like other species of this subgenus, in cylindrical cavities in the earth. Though many specimens were found, I never saw the lid described by authors as closing the aperture of its dwelling. The very singular formation of its abdomen, which is as hard as leather behind, and which forms a perfect circle, induces me to believe that it closes with that part, its dwelling instead of with a lid, when in danger." Along with drawings of the animal, Hentz included a sketch of "the hole in which it resides," a simple, circular aperture in the ground, unadorned by semblance of lid, turret, or silken structure of any kind. Did Hentz actually see the entrance to a burrow? Did he draw upon nature or his imagination as a model for this sketch? We know that he never saw a lid and we can only surmise as to whether or not he saw the entrance.

In his monumental work, *American Spiders and Their Spinningwork,* Reverend Henry C. McCook treated the natural history of spiders in great detail. His chapter, "Enemies and Their Influence on Habit," speculates further on *Cyclocosmia.* Led on by the singular "adaptation" of the abdomen, and encouraged by the work of Hentz and Ausserer, McCook sees in this hard disk "one of the most curious examples of relation of structure to enemies, or perhaps of the reaction of hostile environment and agents upon structure." Relying solely upon Hentz for his information, but cautiously warning that Hentz's conjectures need confirmation, he agrees that it is not improbable that *truncata* uses his abdomen as a door. He further appends a beautiful sketch of the spider in this imagined position, and remarks: ". . . and one may imagine the intellectual confusion of a pursuing enemy, which finds its prey suddenly disappearing within a hole in the ground, but which, when investigated, presents nothing but a level surface where certainly a hole ought to have been."

For many years no specimens of *Cyclocosmia* were seen by spider students but more recently colonies of *truncata* have been found in northern Georgia, Tennessee, and Alabama. Credit for discovery of *Cyclocosmia torreya* belongs to Dr. H. K. Wallace of the University of Florida, who found the well-hidden burrows in the bottom of Torreya Ravine. *Cyclocosmia torreya* seems to prefer a rather steep slope in a shady, cool, somewhat damp location. The first burrows found were in a vertical bank protected by the overhanging roots of a large tree, a situation characteristic of the ravines in Torreya Park, where small streams have been actively eroding their courses. These exposed red and yellow, sandy clay surfaces are partially covered with mosses and liverworts. The burrows are straight, cylindrical, and almost vertical in every instance. They are enlarged for two-thirds their upper length, then narrow abruptly until they are exactly the diameter of the hard abdominal disk of the occupant. Specimens are usually found headfirst in the bottoms of the burrows, presenting their armor plate to the intruder. In this position they fit the cylindrical cavity so nicely, and they hold on with their claws so tenaciously, that it is necessary to dig the earth away in order to extricate them without injury. When disturbed, some back up their burrows to where there is room for them to turn and present their fangs.

The burrow of *Cyclocosmia* is covered by a hinged trap door, which is similar in shape to that of *Ummidia* but much thinner and quite flexible, thus belonging to the wafer-type. Most of the doors appear to be located in and under leafmold on the sides of the banks, a circumstance that makes them difficult to locate.

It has now been established that *Cyclocosmia* is simply another trap-door spider, but an extraordinary one; what, therefore, can we conclude regarding the previous interpretations by older students of the use of its abdomen, interpretations that have persisted even into recent books and papers? Obviously, it is disproved that the spider closes the burrow with its abdomen, which is not large enough to plug the opening. *Cyclocosmia* seemingly has two lines of defense against enemies: its well-hidden surface door and its ability to run down to the bottom of its burrow and completely plug the tube.

The protective devices (Fig. 4) of the trap-door spiders herein considered may be briefly reviewed as follows: *Ummidia, Bothriocyrtum*, and our *Eucteniza rex* of southern Texas, rely upon a fortress guarded by a heavy cork door, which they hold shut with surprising strength. *Myrmekiaphila, Actinoxia*, and *Aptostichus* build a weak, flexible cover that serves only to keep out rain, but is well camouflaged; they depend upon the deception of the concealed side chamber deep in their burrow. *Cyclocosmia* carefully hides the wafer door to its nest, and to intruders presents its tough body armor as a shield.

Sheet-Web Tarantulas. The spiders of the family *Dipluridae* have followed a course in their development quite different from either the trap-door spiders or the tarantulas. They spin a silken funnel in a crevice, under rocks, or in thick vegetable growth, and then continue the silk over the ground as an expansive sheet. The spider hides in the funnel, and waits for insects to fall upon the funnel or become entangled in the sheet webbing, whereupon it rushes out and captures its prey. This type of web is called a sheet web; it is the same in general plan as those spun by the grass spiders, by some wolf spiders, and by one group of atypical tarantulas. The spiders that use this device for capturing insects are usually agile hunters, which can rush to the location of their prey with great speed. Their movement on the flat sheet has, in a nice comparison, been likened to a skier gliding over the top of the snow, whereas the bulky insects make headway on the yielding silk like a man walking through heavy drifts.

The sheet-web tarantulas are in many ways specialized spiders. They have developed the best eyesight of all the mygalomorph spiders. Their bodies are quite long and flat, and the tarsi of their long legs are provided with an unpaired claw, as in the trap-door spiders. Since much of their prey drops on their webs during the day, they hunt equally well during the day and night. Their spinnerets are frequently greatly elongated and widely separated. The terminal segments of their long lateral spinnerets are provided with many small spools, from which can be spun a wide sheet of silk when these organs are moved from side to side. Except for two exotic genera, *Hexathele* of New Zealand and *Scotinoecus* of Chile, which have six spinnerets, only four spinnerets are present.

Most of the diplurids live in the tropics where large species and great sheet webs are conspicuous objects. In the United States only three genera occur and though not notable for size they are otherwise interesting. *Microhexura* is of particular interest because they are among the smallest of all tarantulas, averaging about one-eighth inch in length. These tiny creatures carry their egg sacs around with them in their chelicerae, held beneath the body between the front legs much as do the fisher spiders. Three different species are now known, one from the high mountains of North Carolina and Tennessee and the other two species from mountains in Washington and Idaho. They spin their small sheet webs under pieces of bark, in decaying wood, under logs, and deep debris, in moist deciduous woods, or fairly dense coniferous forests.

The species of *Euagrus* are considerably larger than *Microhexura*, running half an inch in length, and even longer in the tropics. Several species occur from southeastern Texas across into southern Arizona. All our species build thin webs on the ground, especially in rocky situations, and the funnels are hidden away in crevices or under stones. They are usually pale yellow or light brown, but the many larger species from Mexico are dark brown or black. Three species from Mexican caves lack eyes completely. The males are remarkable in having the tibia of the second legs much swollen and armed near the middle with a heavy spur containing clasping spines which aids in holding the female during pairing. On the prolateral face of the second femur and facing it on the first femur are long patches of tiny recurved spinules that, when rubbed together, are likely sound producing organs.

The third diplurid genus in the United States is *Calisoga* with two species known exclusively from the Coast Ranges and Sierras, respectively, of California;

these were erroneously referred to the exotic genus *Brachythele*. *Calisoga longitarsis* is a large spider, measuring from one to two full inches in body length, and equal in size to many tarantulas. It is said to dig a deep burrow into the soil of banks and line only the outer part with silk, often spinning up the entrance with scattered lines. This superficial outer web probably represents an undeveloped sheet web and is similar to that spun by the Australian *Atrax* and other relatives. The smaller males wander in the summer in search of females and the eggs are probably produced in the fall. The species from the Sierras, *Calisoga theveneti*, differs little from *Calisoga longitarsis*. These large diplurids are quite belligerent, defend themselves with great spirit, and in this temperament are similar to the Australian *Atrax*, a genus noted for its dangerous venom.

Tarantulas. Largest of all spiders are the immense hairy creatures of the family *Theraphosidae*, which Americans call "tarantulas." Although these mygalomorphs have nothing in common with the wolf spider of southern Europe, which truly deserves the name "tarantula," they have so completely usurped this appellation that any attempt to change it would be futile. In most of Spanish America, the covering of hairs on the legs and bodies of these spiders has earned them the name of *arañas peludas*—"hairy spiders." Not inappropriately, they are dubbed by the Brazilians *carangueigeiras*, because of the long bony legs—especially of the males (Plate VII)—and their stance and gait give them a superficial resemblance to crabs. In Argentina, many are called *arañas pollitos* or little chicken spiders. In Mexico, native Indian names have largely been displaced by "tarantula," which is applied to almost any large spider. But in Central America, where these creatures are reputed to be dangerous to horses, they are still called *arañas de caballo*, or *matacaballos*. Outside the Americas, the tarantulas are widely referred to as "mygales," or "bird spiders." This latter name is inappropriate and largely inaccurate, because most of the species are ground-loving types and have little opportunity to attack birds on the ground or in trees.

No matter by what name the tarantulas are known, they excite the imagination because of their great size and notoriety (Plates 3, 4). In the steaming jungles of northern South America live the largest and bulkiest members of the whole tribe, enormous spiders that have no peers for size anywhere else in the world. A male *Theraphosa* from Montagne la Gabrielle, French Guiana, measured three inches from the front edge of the chelicerae to the end of the abdomen, and had a leg span when fully extended of ten inches. An enormous female perhaps of the genus *Lasiodora*, from Manaos, Brazil, the bulkiest tarantula I have ever seen, had a body three and one-half inches long, and measured nine and one-half inches with the legs extended. Quite handsome in her clothing of fine brown hairs, she weighed almost three ounces. Some of these giants are said to attain full maturity in three or four years.

Our United States species are pygmies by comparison. A full-grown male of *Dugesiella* from Arkansas was found by W. J. Baerg to weigh a little less than one-half ounce. The greatest total length of the carapace and abdomen of this specimen was about two inches. A representative female of the same species closely approximated the male in weight and body length. Large females of our species often weigh as much as two-thirds of an ounce after they have been well fed. The long legs of our southwestern males span six or even seven inches.

Owing to their formidable appearance, the tarantulas have acquired the reputation of being dangerous. This is a reputation they do not live up to either in belligerence or in the virulence of their bite. For the most part, they are sluggish, phlegmatic creatures, that attack only when goaded to an extreme. Although our species are credited in many accounts with being great runners and jumpers, leaping is not their speciality, and they ordinarily leap to strike over only a few inches. In point of fact, they make fine pets, and some quickly become so tame that they can be picked up and handled with ease. These animals have become favorite house pets (Plates 3, 4) and species from many parts of the world are imported by pet shops, and many sold at exorbitant prices. Species from Texas and Arizona are among the favorite native ones that are being exploited, much to the consternation of ecologists, who deplore the depletion of once large colonies in some localities. The venom of most tarantulas seems to have little harmful effect on man, but the powerful chelicerae of large species are capable of producing painful wounds.

About thirty species of tarantulas live within the limits of the United States, and for the most part in the arid Southwest. The absence of tarantulas from Florida and the southeastern states is rather surprising since that area is seemingly ideal for such hairy spiders. Only a few miles away in the Bahamas and the West Indies is found a rich fauna of tarantulas; these belong to a quite different group than our southwestern species. The range of our species is eastward only to the Mississippi River. To the north they occur in Missouri and across into Colorado, Utah, southwestern Idaho, and adjacent Oregon, and in California up to the top of the central valley and foothills, and in the San Francisco region. South of this general distribution pattern they are represented by many distinctive species, and by many more southward into Mexico.

Tarantulas abound in the tropics and there have developed many interesting types. A few of them have become arboreal and move over the surface of trees with great facility, aided by adhesive tarsal pads. Some species nest in bromeliads and other stations far above the ground. The generic name *Avicularia* was prompted by one of these arboreal types that occasionally prey on birds. Even the ground-loving types of tarantulas are good climbers, since their tarsi are also provided with adhesive hairs that enable some of them to climb a vertical pane of glass with ease. The tarantulas of our American Southwest, on the other hand, are more restricted in habit. They dig their own burrows (Plate VIII), starting them when they are spiderlings, and live in them for many years. Once they have established a burrow, they usually live there during their whole life. The area in which they hunt is small, only a few yards on each side of their burrow, and they rush back into their tunnels at the slightest disturbance. Rarely do they live in regions of dense forest or heavy undergrowth, preferring open areas on hillsides, mixed desert growth, or the fringe of cultivated land. The burrow usually has a loose webbing at the entrance, spun there after the night's hunt and indicating that the spider is at home. During the winter months the opening may be plugged with silk, leaves, and soil, and in some instances a little mound of earth surmounts it.

All spiders need water, and tarantulas are no exception. Indeed, Baerg attributed the complete disappearance of a large colony of Mexican tarantulas near Tlahualilo, Durango, to a drop in the normal rainfall from nine to three inches. On the other hand, small quantities of water poured into the burrow will often bring the spider rushing out into the open—a procedure that affords an easy means of collecting them. The tarantulas in the rain forests of the American tropics frequently live above the ground, and after heavy rains may be seen wandering around in the open. Aversion for water—and this describes most tarantulas well—may have inspired some tarantulas to become arboreal, and thus escape regular deluges that they might have experienced on the ground or below the surface.

Tarantula burrows (Plate VIII) are often tunnels under large stones. Within spacious confines the mother spider spins a large sheet, upon which she deposits her large eggs. She then covers them over with a second silken sheet and binds the edges together to form a flabby bag. For six or seven weeks she watches over this sac, occasionally bringing it to the entrance of the burrow to warm it in the direct sunlight, until finally the babies emerge. The spiderlings are gregarious, and often remain in the burrow for days or even weeks; eventually they disperse by walking out of the hole and moving in all directions. Since they are much too large to balloon away, they settle down in the general neighborhood of the burrow, hiding under chips and stones for a time and then making their own tiny burrows in the ground. As with all spiders, there is a tremendous mortality in the young stages: from each sac perhaps only a pair of tarantulas reach maturity.

Adulthood for the spiderlings is very far in the future since eight to ten years are usually required for either sex to become sexually mature. The females and the immature males live in similar burrows in the ground, and they remain virtually indistinguishable until after the last molt, at which time some are surprisingly revealed as males. Many large spiders of this group kept in cages for years and given some common feminine name will molt to reveal the longer legs and brighter colors of the male. Typical males are much darker than the brownish females, and have an abdomen set with rusty red hairs.

Their final transformation gives the males an entirely different outlook on life. Whereas they have been content for years to live in a dark burrow, they now desert it and wander over the countryside in search of mates. This searching for mates occurs from June to December in the Southwest, and during this period they may be seen crossing the highways, frequently in considerable numbers. Almost all of the walking tarantulas are males and are seen only during the mating season. Few survive the year in which they become mature; many die a natural death, others are killed by females during courtship or after mating. Thus, the males of our southwestern tarantulas live a total of about ten years. It is quite different with the females who go on living many more years.

Living to a ripe old age is quite an accomplishment inasmuch as tarantulas are plagued by many enemies. In the Southwest, various digging animals such as skunks, coatimundis, javelinas, and others, dig out the tarantulas and use them for food. The young tarantulas are preyed upon by birds, lizards, frogs, and toads, and some snakes find them quite suitable dietetically. Insidious enemies are the small-headed flies of the family *Acroceridae*, known to confine their attentions exclusively to spiders, in the bodies of which they develop as voracious maggots. The species of *Pepsis*, giant metallic blue or greenish wasps with rusty wings, specialize in tarantulas (Plate IX); those few that occur outside the range of these spiders probably must prey on large insects. Preferred prey because of their greater bulk, the female tarantulas offer a far more generous supply of nutritional food to this predator than do the male tarantulas. The long legs of the male seem to give him some degree of safety, and when he elevates his body high on his legs, the "tarantula hawk" has such difficulty in stinging him that she may abandon her efforts.

Pepsis forages over the ground alert for any female tarantula venturing out during the day and is able to move into the burrow to secure her game. The details of a struggle, quite as unequal as in the case of the trap-door spiders, are given by Alexander Petrunkevitch:

> The *Pepsis* comes deliberately to the tarantula on the side of the cage and drives her down to the ground. The next moment she closes in on her victim in the manner already described, and bending her abdomen under the venter of the tarantula, introduces the sting between the third and fourth right coxae, close to the sternum. The tarantula struggles violently and rolls with the *Pepsis* over and over on the ground. After a few struggles, the *Pepsis* lets go her hold on the tarantula, walks off a couple of paces, turns and comes directly toward the jaws of the tarantula. Without the slightest hesitation,

she slips under the tarantula, which raises as high as she can on all her legs. The *Pepsis* grabs the fourth leg with her mandibles. The tarantula tries to bite her enemy, but the *Pepsis* holds her off by pressing her feet against the feet of the spider, while at the same time continuing her hold on the fourth leg with her mandibles. Meanwhile, she bends her abdomen and searches for the place to pierce with her sting. Now she finds it. It is the same place as in the first specimen, that is, the articulation membrane between maxilla, first leg, sternum, and lip. In a few seconds the tarantula is paralyzed. The position of the two is very remarkable. The tarantula sits in her normal way, but the *Pepsis* lies on her right side, head toward the posterior end of the tarantula, sting in the place mentioned. After at least half a minute, the *Pepsis* withdraws her sting and walks off. The tarantula remains motionless. Presently one leg of the tarantula moves. The *Pepsis* returns, climbs on the tarantula, inserts her sting between the sternum and the third coxa and holds it there for about a minute.

All that remains is the transport of the heavy spider, often weighing eight or ten times as much as the wasp, to its grave, which may have already been dug or which may be dug after the capture. Once the victim is within the prepared cavity, an egg is deposited on its abdomen and the burrow sealed up. The paralyzed spider provides a fresh supply of food for the larva of the wasp and may remain alive for months. Only rarely does a spider recover from the effects of the venom and escape its fate.

The tarantula reacts to enemies in various ways. By throwing itself back on its haunches and elevating its head to expose formidable fangs, it assumes a defensive attitude that may frighten away timid adversaries. If a tormentor persists in goading the spider, the spider often elevates its abdomen and, working its hind legs rapidly, scrapes loose a small cloud of extremely fine abdominal hairs. When these hairs come in contact with the mucous membranes of the eyes or nose of mammals or man, a very disagreeable urtication results, which persists for some time. In discouraging small mammals, this may be effective, perhaps allowing the spider to escape while the aggressor is still partially blinded. American tarantulas usually have a distinctive patch of such urticating hairs on top of the abdomen, which when put to full use results in a bald spot devoid of the poisonous hairs. After the next molt the spider is provided with a full new covering of hairs and setae. Unfortunately, this protective device can have no effect on the solitary wasps which are most important as predators.

The body hairs of tarantulas have long been known to have an urticating effect on the skin of man; in allergic individuals they sometimes produce distressing symptoms. It is probable that a toxic substance is present on the hairs, and that the effect is not entirely mechanical. Support for this view is seen in the fact that alcohol in which these spiders have been preserved is capable of producing the characteristic itching and stinging. *Brachypelma emilia*, a handsome Mexican species, is capable of producing severe swelling and stinging with its urticating hairs.

Because all tarantulas of the United States are ground forms, their food consists largely of animals available in their restricted hunting areas. Beetles and grasshoppers are most frequently captured, but many other kinds of insects, and crawling creatures as sow bugs, some kinds of millipedes, and other spiders, fall to their lot. It is well known that our species will kill and eat frogs, toads, mice, and lizards in captivity, and it is reported that occasionally these small vertebrates are captured in natural surroundings. During the summer the tarantula catches and eats insects almost every night, frequently gorging itself. On the other hand, long periods of fasting seem to have little effect on these spiders. In order to ascertain just how long they could go without food, Baerg kept several tarantulas supplied only with water. One of the females lived two years and four months without food, and other females almost matched this record.

Though the belief is more widely held than is justified, tarantulas have long been known to capture and feed on small birds. The first record of this behavior was published in 1705 by the Swiss naturalist Maria Sibylla Merian in her *Metamorphosis Insectorum Surinamensium*. A fine color plate shows one of the South American mygales in the act of feeding on a hummingbird. The spider, a great brown creature said to belong to the genus *Avicularia*, has its fangs embedded in the breast of the gaily colored bird, which has been struck on its nest. Mme. Merian's report (which was received with considerable skepticism, since it was not believed at the time that any vertebrates could be consumed by spiders) was later followed by many claims that birds, lizards, and other animals were habitual prey of the great tarantulas and even of other smaller species. Corroboration of the early stories came in 1863 from H. W. Bates, in his book *The Naturalist on the River Amazon*. This talented observer actually saw the capture and killing of one of two birds that were attacked, and very accurately depicted the spider in the act of feeding on it. Since that time, the debate has been concerned with the capability of the spider to actually make use of the body of the vertebrate as food, not with its ability to capture it.

That a powerful, predaceous animal, armed with strong fangs and potent venom, can kill a bird, a mammal, a snake, or a lizard is not an astonishing thing. The arboreal tarantula cannot differentiate between a bird or a large insect, and makes its capture in exactly the same manner—by springing upon it and striking it down with its fangs. Spiders predigest their food by flooding the victim with secretions from the maxillary and other glands, softening the tissue so that the liquid nutrients can be sucked into the mouth. The powerful buccal secretions are known to have a digestive effect on meat, so it is not strange that even the bodies of vertebrates can be taken through the small mouth opening. A tarantula can reduce the fat body and wings of a large saturniid moth to an insignificant blob, and do so thorough a job of it that one wonders if chitinous outer parts were not absorbed along with the softer portions. In can consume the bulk of a fat mouse or the body of a small rattlesnake in the same way, feeding on the gruesome corpse for many hours.

In the United States the lessened opportunity to capture small vertebrates has kept our tarantulas largely insect eaters—a quite different situation from that in Brazil, where the giant species of *Grammostola* and *Lasiodora* are believed to kill and feed on frogs, lizards, and small snakes in their natural surroundings. In captivity, these large spiders definitely preferred such small cold-blooded animals, and would generally pay little attention to various insects offered as food. While experimenting on spider venoms, Dr. Brazil and Dr. Vellard of São Paulo kept fifty of the tarantulas in good health for eighteen months on a diet of frogs, lizards, and snakes. Small rattlesnakes and the venomous *Bothrops* were killed and eaten as readily as any other kind of snake.

When a *Grammostola* and a young snake are put in a cage together, the spider tries to catch the snake by the head, and will hold on it spite of all efforts of the snake to shake it off. After a minute or two the spider's poison begins to take effect and the snake becomes quiet. Beginning at the head, the spider crushes the snake with its mandibles and feeds upon the soft parts, sometimes taking twenty-four hours or more to suck the whole animal, leaving the remains in a shapeless mass.

One of the interesting bits of folklore prevalent in southern Mexico and Central America is the legend of the *matacaballo*. For many years it has been a general belief that tarantulas bite the fetlocks of mules and horses and cause the loss of the hoof. According to the story, the spider hunts out the sleeping animal at night and takes a narrow strip of hair from above the hoof for its nest building, using an acidlike secretion to make the hair slough off more easily. The site of the injury then becomes inflamed, infection occurs, and the hoof may eventually be lost. In another version, all goes well unless the spider is disturbed and bites the hoof. In order to prevent hair clipping by the *matacaballo*, the natives run their animals through a footbath of water covered with an inch of crude oil. It is said that the tarantulas do not like the oil-covered hair, so the animals gain temporary immunity from the presumed scourge.

It is now known that this often fatal disease is actually caused by a bacillus that is prevalent in the soils of Central America. During the rainy season the skin of the horse's hoof becomes chapped and the bacillus is able to enter through small abrasions. Needless to say, tarantulas use only their own white silk for their nests.

THE ATYPICAL TARANTULAS

One of the two principal branches of the *Mygalomorphae* has culminated in the *Atypidae*, the purse-web spiders; they are the namesakes of the series known as "the atypical tarantulas." This series and the typical tarantulas as well have been derived from forebears similar to our living liphistiids (family *Liphistiidae*), which have changed little since the late Paleozoic era and are the last remnant of an ancient group that failed to alter its form to cope with an altered environment. Near relatives of the atypids are the sheet-web atypical tarantulas (family *Mecicobothriidae*), the antrodiaetids and relatives (family *Antrodiaetidae*), and the above-mentioned purse-web spiders. The atypical tarantulas (Fig. 5) have paralleled in their development the other principal branch of the suborder, the tarantulas and trap-door spiders, and have matched rather closely their handiwork in silk.

An outstanding characteristic of the atypical tarantulas is the presence of a small number of dorsal sclerites at the base of the abdomen. Although of smaller size these sclerites differ little from those found in the family *Liphistiidae*. The atypical tarantulas are of moderate size, few of them exceeding an inch in length, and in general form and appearance they resemble the typical trap-door spiders. Most of them are accomplished burrowers but only the antrodiaetids and their close relatives have the chelicerae fitted with a rake of coarse teeth for digging. The unpaired claw is present on the tarsi, but no claw tufts or tarsal brushes have been developed. The atypical tarantulas long ago lost the anterior median pair of spinnerets still retained by the modern liphistiids. The persistence of the anterior lateral pair of spinnerets is noteworthy, since it is retained elsewhere among the mygalomorph spiders only in two primitive genera of the family *Dipluridae*. The anterior lateral spinnerets are two-segmented and functional in the four known genera of the *Mecicobothriidae*. They are two-segmented and functional in *Aliatypus*, one-segmented and small or reduced to vestiges in *Atypoides*, and completely aborted in *Antrodiaetus*.

The atypical tarantulas retained some features that were lost by the typical tarantulas, but were retained and exploited by the true spiders. The male palpus is provided with a conductor of the embolus—a shield to protect the tube conveying the semen—which is found in almost none of the typical tarantulas and is apparently the homologue of that found in the true spiders. The epigyna of the females all agree in having four primary seminal pouches, whereas in almost all of the higher *Mygalomorphae* and in all but a few of the true spiders there are only two.

The atypical tarantulas are hardy creatures that live much farther north in the United States than any of the typical tarantulas. Some of the antrodiaetid tarantulas are common in our southeastern states and rare northward into New Jersey

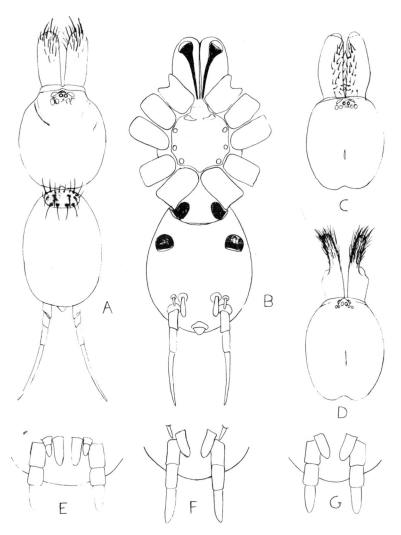

Fig. 5. The atypoid mygalomorphs. A. Dorsal view of male *Hexura picea*, legs omitted. B. Ventral view of female *Hexura picea*, legs omitted. C. Carapace of male *Megahexura fulva*, with spinose chelicerae. D. Carapace of *Atypoides riversi*, with hirsute apophyses on chelicerae. E. Spinnerets of *Aliatypus californicus*. F. Spinnerets of *Atypoides riversi*. G. Spinnerets of *Antrodiaetus unicolor*.

and New York; in our northwestern states they extend into British Columbia and Alberta, and even into adjacent Alaska. In Europe *Atypus* is found in England, and the same species occurs in Denmark, a location that would place it above the fiftieth parallel north. In the United States *Atypus* is uncommon in the north but has been taken in Massachusetts and Wisconsin, and at least once in Ontario, Canada, stations all well above the fortieth parallel.

Sheet-Web Atypical Tarantulas. The family *Mecicobothriidae* parallels very closely the sheet-web tarantulas of the family *Dipluridae;* indeed, until quite recently they were placed in that family. Since both of these families are placed near the base of their lines, it is possible that the diplurids were derived from prototypes of the mecicobothriids and that they are actually cousins. The hind spinnerets of both of these spider families are greatly elongated (particularly the terminal segment, which is flexible) and rather widely spaced; this is probably an adaptation for spinning the sheet webs. We find *Hexura* running over a sheet web similar to that of *Euagrus* (another typical diplurid). The family *Mecicobothriidae* was established on the basis of a single male specimen from Tandil, Argentina, which has long been lost and of which no specimens have been taken since or are known to be deposited in any world museum. The family is otherwise known only from three genera found in the western United States.

The genus *Megahexura* (Fig. 5, C) is represented by a single, medium-sized species, *fulva*, with females sometimes three-fourths of an inch long; these spiders are widespread in California. Two separate tergal plates lie above the pedicel of the abdomen. This genus differs from *Hexura* in having the anterior lateral spinnerets two-segmented, and the anterior slender tibia of the male unspined, without clasping spines or modifications.

The genus *Hexura* is represented by the classic species *Hexura picea* (Fig. 5, A, B) smaller than *fulva* and rarely exceeding one-fourth of an inch in length. This dusky-brown spider, widespread in Oregon and Washington, lives under leaves, trash, and pieces of wood or bark on the ground in pinewoods, where it builds a loose sheet web in which it stays and over which it hunts. The abdomen has above the base a brown scutum composed of two coalesced tergal plates. The anterior lateral spinnerets are one-segmented; in a still undescribed species from southwestern Oregon this pair of spinnerets is aborted. The anterior tibiae of the males are moderately thickened and bear a patch of clasping spines that hold the female during mating.

The third American genus, still undescribed, consists of four tiny species, only about one-eighth of an inch in length, that parallel in general appearance the pygmies of the diplurid genus *Microhexura.* Two of these species are found in Arizona and the other two species are known from southern California and adjacent Baja California. This genus, which will be named *Hexurella* because of the small size of the species, has several unusual features. The posterior lateral spinnerets are four-segmented, a condition found in our *Atypus abboti* of Florida. Two small, inconspicuous tergites occur above the base of the abdomen. The first leg of the males is moderately thickened and sometimes bears on the femur or tibia a patch of clasping spines.

The Antrodiaetids. The family *Antrodiaetidae*, one of central position among the atypical tarantulas, is exclusively North American except for two species found in Japan. The family has recently been studied by Dr. Frederick A. Coyle, who has presented a wealth of new biological, ecological, and systematic information for the twenty-five American species. An important feature of this group is the possession of a distinct rake on the chelicerae, which marks them as expert burrowers. For this reason they were long placed among the true trap-door spiders of the family *Ctenizidae*, a group they resemble closely in habitus and have paralleled in their development. In its two-segmented anterior lateral spinnerets *Aliatypus* (Fig. 5, E) would seem to be more generalized but in other features, and especially in its construction of a trap-door for its burrow, it appears to be specialized. In *Antrodiaetus* the anterior lateral spinnerets have been lost, and in the related *Atypoides* these spinnerets are small or reduced to vestiges (Fig. 5, F, G). The presence of two, three, or four well-marked tergites at the base of the abdomen in both sexes is invariable; these are strikingly large and distinct, set with rows of transverse setae much as in the liphistiids.

The turret spider *Atypoides riversi* is the best known of our three species and lives in the Coast Ranges of the central and northern part of California and across from this in the Sierras. It is found in abundance along shaded streams and in thickets in the San Francisco Bay region, where its turrets are well-known objects. These are ordinarily stiffened, silken turrets open at the top, lacking a formal trap door, but with a collapsible collar that closes the opening, much like that of *Antrodiaetus*. The burrow is very long, usually inclined, and is lined completely and rather heavily with white silk. The aerial portion may be only a short chimney, but quite often there is a long tube, which, penetrating thick grass, moss, or debris, finally terminates in the expanded white lip of the turret. The spider takes whatever building materials are handy—leaves, small twigs, moss, bits of lichen, pine needles—and fastens them on the outside of the silken collar. Often most ingeniously constructed, the turret provided an excellent lookout for the spider, which sits in the entrance at dusk and catches the insects that come within its reach.

The turret spiders are about half an inch long, with yellowish brown carapaces and darker brown or purplish abdomens. A remarkable feature of the male is the presence of a long, projecting process on each chelicera (Fig. 5, D), which is probably used for holding the female in some way while mating. The tiny anterior lateral spinnerets are composed of a single joint and, judging from their reduced size and lack of spinning openings, are rapidly being aborted. The median groove on the carapace is a linear impression. Another related species, named *gertschi*, is somewhat paler and lives in the northern Sierras of California and into the Cascade Mountains of southern Oregon. A third species, *hadros*, with remarkably disjunct distribution, has been found in southern Illinois and Missouri.

The most familiar and widespread of the antrodiaetids are the species of *Antrodiaetus*, the type genus now with eleven species in North America and two distinctive species in Japan. The short anterior lateral spinnerets, greatly reduced in size in *Atypoides*, have here been completely lost. The carapace has the median groove present as a longitudinal depression. In the males the dorsum of the abdomen is armed with three distinct tergites, and in the females one or more tergites are present. The chelicera of the males is armed with a prominent tubercle set with black setae, which probably aids in some way during the mating in conjunction with patches of setae on the legs of the males.

These spiders live in burrows, which may descend a foot or more into the soil, and which often have prominent bends. The upper part of the burrow is usually well lined with silk; in western species the opening is often concealed under stones or hidden in debris. As a result of their secretive habits and well-hidden burrows, females are more difficult to find and usually have to be found by digging. The adult males desert their burrows and wander about in search of females during the summer and fall months. Each area of our country has representatives of the genus *Antrodiaetus: unicolor*, a common and somewhat variable species ranges in the foothills and mountains of the central and southern Appalachians west to the Ozarks; *robustus*, a related species has a narrow range from Maryland into adjacent states and is the species most often seen in the Washington, D. C. area; two species, *lincolnianus* and *stygius*, occur in the grassland states from Nebraska to northern Arkansas; *apachecus* lives in the mountains of Arizona and New Mexico; and the remaining six species are found in the northwestern states. Of these latter can be mentioned the blackish species *pacificus*, which ranges in mesic coastal situations from San Francisco to southern Alaska, and the pale Great Basin representative, *montanus*, which favors sagebrush and foothill country.

In 1886 George F. Atkinson studied the species *unicolor* in North Carolina and, because of the singular means by which it closed its burrow, called it a "folding-door tarantula." As he described it, there are two equal doors, each forming a half circle, which hang on semicircular hinges; when closed, they meet in a straight line over the middle of the hole. Each night the spider throws open its burrow, and each morning closes the doors. This description of the closing flap is innacurate and only vaguely conforms to that of a folding door. The opening of the burrow is surmounted by a collapsible collar of silk, into which is incorporated soil and plant pieces that camouflage it. The spider closes the door by seizing it bilaterally and pulling it into the opening; the formal appearance of a folding door is only approximated.

On the method of capturing its prey, Atkinson had the following to say:

One evening I placed several ants in the jar containing the nest. When an ant approached, so near the door as to send a communication to the spider of its presence, the spider sprang to the entrance, caught a door with the anterior legs on either side, and pulled them nearly together, so that there was just space enough left for it to see the ant when it crossed the opening. When this happened, the spider threw the doors wide open, caught the ant, and in the twinkling of an eye had dropped back to the bottom of the tube with its game. This I saw repeated several times during the months of January and February.

In many ways the third genus of this family, *Aliatypus*, is the most derivative of all in having perfected a trap-door closing of its burrow. Except for a single disjunct species from mixed deciduous, montane stations in Arizona (*isolatus*), the remaining ten species are exclusively Californian, and among the commonest mygalomorphs in the state. The first described species, *Aliatypus californicus*, common in the mountains and foothills of the San Francisco Bay region and across the valley in the Sierra Nevada Mountains, represents the genus. The burrow is comparatively long and either goes straight down into the compact soil or is provided with pronounced bends. The silk lining is quite thin, but thickens around the opening, which is covered with a formal trap door of the wafer type. The burrows are usually found along roadside banks and streams, where the spider seems to prefer exposed soil only thinly covered with vegetation.

The female *californicus* resembles the turret spider, but has a somewhat broader carapace marked by a round median groove. The male resembles the female quite closely and completely lacks a distinct spur on the chelicerae such as is present in *Atypoides*. The male palpi are thin appendages as long or longer than the first pair of legs. A most interesting feature of this genus is that the anterior lateral spinnerets (Fig. 5, E) are nearly equal in size to the posterior median, and are also two-segmented. Well-developed spigots show that they are functional appendages—a fact that marks them as being among the most generalized of all mygalomorph spinnerets, except those of the *Liphistiidae*. Since they are two-segmented, we can state with confidence that they are definitely the anterior lateral pair, and thus corroborate on direct evidence what has been the presumption of the majority of araneologists. The remaining nine species of *Aliatypus*, all recently described, are distributed over much of California, and occur in wet redwood forests, cool montane forests of the Sierras, and also in hot areas of sagebrush and scrub plants. The adaptive radiation of this genus parallels that of the true trap-door genus *Aptostichus* in California.

Purse-Web Spiders. In the low hammocks of Georgia and Florida lives one of the most remarkable members of the tarantula fauna. It has received the common name of "purse-web" spider from the resemblance its web bears to the silken purses so much favored by ladies over a century ago. In 1792 John Abbot, eminent entomologist and artist of Savannah, Georgia, first described the tubes of the species that bears his name:

> This singular species makes a web like a money purse to the roots of large trees in the hammocks or swamps, five or six inches out of the ground, fastened to the tree, the other end in the ground about the same depth or deeper. To the bottom of that part in the ground the spider retreats. I imagine that they come out and seek their food by night as I never observed one out of its web. In November their young ones in vast numbers cover the abdomen of the female and the abdomen then appears very shrunk. The male is the smallest, but has the longest nippers. Taken in March and is not common.

Atypus abboti (Plate 7) digs a deep burrow in the soil at the foot of a tree. This it lines with silk, then prolongs the silken lining up the side of the tree. The aerial tube is securely fastened to the bark by threads, and in full-grown females is about ten inches long and three-fourths of an inch wide. Smaller specimens spin correspondingly smaller tubes, which are almost invariably placed upright against a tree. The top of the tube is open, but the silk is so flattened and pressed together that the natural opening seems to be closed. An even covering of sand and other fine material serves to color and darken the white silk and make it less conspicuous. In Florida the tubes are most often found attached to sweet gums, oaks, and magnolias in deep forest where the soil is damp and rich in organic material; these tubes also have been observed in dry woods where the sandy soil has little or no covering of humus. The tubes of *Atypus bicolor* (Plate 6), our largest species which occurs sparsely even north to Connecticut and Rhode Island but is more abundant from Maryland south to eastern Texas and Florida, are often eighteen inches long. This species lives in mesophytic woods and swampy situations. Near Quincy, Florida, I found them abundant in deep woods near a small stream.

The tube of *Atypus* takes form in a characteristic manner. The female spins a small, horizontal funnel or cell on the surface of the soil, and from this base works both upward to lay out the aerial tube, and downward into the soil. The funnel is pierced above, and a two-inch section of vertical tube is set up against a tree. This design is accomplished by laying down many single lines and spinning the whole together into a strong fabric. The spider then begins excavating and spinning the subterranean part of her habitation. She molds the soil into small pellets, which she disposes of through the opening at the top of the aerial web. The covering of debris over the surface of the tube comes, surprisingly, from within the burrow—instead of being laid on from the outside: the sand and small particles are pressed outward through the web until the whole surface is evenly covered. After the first section of aerial tube is completed, another length is spun and coated with sand. Thus by sections the web moves up the side of the tree, until it attains the full length for the species. Like an iceberg, the finished tube penetrates the ground much farther than the length of its visible, aerial portion. It is heavily lined with silk, which becomes stronger day by day as the spinnerets constantly lay down their dense bands.

The purse-web spider remains just inside the subterranean portion of her nest while waiting for prey, but at the slightest notice of a passing insect she moves into the aerial tube. Her course is charted by the movement of the tube, and when the insect crawls over the surface, she rushes to the proper point and strikes her long fangs through the web around or into the body of her prey. Holding her prey until it is completely subdued, she at the same time cuts the tube and pulls the insect inside. A slight rent is left in the silk, which will later be sewed together, and in due time covered over with sand so evenly that no sign of the break will be evident. A tidy housekeeper, *Atypus* when through feeding brings the shrunken remnant to the opening at the top of her tube and casts it out. In the same way, she voids her milky white, liquid fecal material through the opening—with such force that it is shot several inches away.

In June the males become adult and leave their webs to wander about in search of a mate. Until they become fully adult they live in nests that are to all appearances identical to those of the females, and occasionally in season they can still be found in their tubes after just molting to maturity. The mating behavior of our American species has not been described, but it is probably similar to that of the better known European species. When the male finds the tube of a female, he drums upon it with his palpi, and presumably is able to ascertain by the reactions of the female to his drumming, whether he is going to be welcome. After a short period, he cuts open the tube and enters, and the break is repaired by the female. Mating occurs deep in the tube. It is believed that the male lives in the burrow for many months before he dies. The eggs are deposited within the burrow, and hatch during the summer months. The young may stay with the female for long periods, but in most instances they leave the nest in the late summer by walking away a distance and beginning work on tiny tubes of their own. Some have been said to disperse by ballooning a short distance but their heavy bodies and poor threads for flying make this activity a difficult one.

The European species of *Atypus* have habits similar to those of our American representatives, but with this exception: instead of a tube up the side of a tree, they spin a very short aerial cell or tube, about two inches long, which rests on the ground, and is suspended in grasses or attached to stones. The end of this tube is closed and, as in our species, the spider rarely leaves the tube. The leathery tube, rendered less conspicuous by its covering of sand and debris, would seem to afford considerable protection to the spider; this seems to be borne out by the fact that the atypids are largely immune to the attacks of pompilid spider wasps. Because the aerial cell is completely enclosed, and continuous with its subterranean portion, predators must cut through the web to locate the spider.

The purse-web spiders are the most extraordinary of all the atypical tarantulas, as regards both their physical features and their singular habits. The marks of their primitive origin are clearly shown in the presence, above the base of the abdomen in each sex, of a single large tergite. In the reduction of their cardiac ostia to three pairs is clear evidence that their heart has become specialized, or better simplified, at a much faster rate than have other features of the abdomen. The chelicerae, though not provided with a rake for digging, are modified into effective shovels for carrying loads of sand or pellets of soil. The fang is a long, thin spine well designed to pierce the silk and hold the prey.

The species of *Atypus* are found in the north temperate zones of Europe, Japan, and the eastern United States. Species are also reported to occur in Java and Burma of the eastern tropics. Another genus of the same family, *Calommata*, is largely restricted to tropical areas in Africa and the Orient. The several American species of *Atypus* are all confined to the eastern and central portions of the United States, and are most abundant in the extreme southeastern part of their range. There is some doubt that our American species can continue to be retained in the genus *Atypus* along with the somewhat different Eurasian cousins; if not, our species must bear the name *Sphodros.*

The females of all the American species are predominantly brown in color, somewhat shining, and only very sparsely set with covering hairs. The robust body is provided with quite short legs and long chelicerae, and measures about half an inch in length; the largest species *bicolor* is often an inch in length. The males are, in general, similar to the females but have longer legs. In *niger*, a shining black spider, the disparity in size of the sexes is not particularly marked, but in the other species the males are somewhat smaller than their females, and very brightly colored. The abdomen of the male *abboti* (Plate 7) is a beautiful iridescent blue or purple, set against black legs and carapace. The legs of *bicolor* (Plate 6) are carmine and contrast with its deep-black carapace and abdomen, thus making it the most striking of our species. A very distinctive, brownish species of *Atypus* was described a few years ago on the basis of a large series of males and females taken around a swimming pool and near a hedgerow in Delaware County, Pennsylvania. The species, named *Atypus snetsingeri*, is a close relative of *muralis* and other species found in Europe and Asia, but seems not to be the same as any of them. It builds the same kind of purse web as do these Old World species so it is likely that it has been brought into Pennsylvania by accident and is not an endemic American species.

8
The Cribellate Spiders

The true spiders that possess a flat spinning organ close in front of the anterior spinnerets are called "cribellate spiders." This organ, which exists in addition to the usual six spinnerets, is known as the "cribellum." It is the homologue of the anterior median spinnerets and has been retained as a functional spinning organ, whereas in other true spiders it is represented by an inconspicuous vestige.

The cribellum (Fig. 6, C) may be likened to the fused spinning fields of two spinnerets lying nearly flat against the surface of the abdomen, all but the tips of the originally paired fingers having disappeared. The dual character of the organ usually is evident on close examination, which shows an actual division of the field by a longitudinal line or faint ridge, or a pinching at the point of division. The spinning field itself is covered by thousands of tiny spinning openings, which give it a sievelike appearance under magnification, and from which come exceedingly fine threads of viscid silk. The ordinary silken threads of cribellate spiders come from glands opening on spinnerets, as in other spiders. Whenever cribellar silk is combined with the regular threads, the line becomes so characteristic in color and physical appearance that it is called a "hackled band."

Invariably accompanying the cribellum is an accessory comb of hairs called the "calamistrum." This is a line of curved setae, differing somewhat in appearance in the various families, and always found upon the metatarsus of the hind legs (Fig. 6, E). In some males both the cribellum and calamistrum may be reduced in size or essentially aborted. The use of the cribellum and calamistrum together as a spinning and carding apparatus to produce the cribellate thread is essentially the same among all the spiders of the group. Let us consider the method of a typical hackled band weaver of the genus *Amaurobius*.

The cribellum of *Amaurobius* is divided longitudinally by a distinct septum, on each side of which lies a spinning field. Because of this division, the hackled band spun by *Amaurobius* consists of two ribbons instead of the one band usually found in the cribellates that have obliterated the limits between the two spinneret fields. The two ribbons are borne by two strands of dry silk presumed

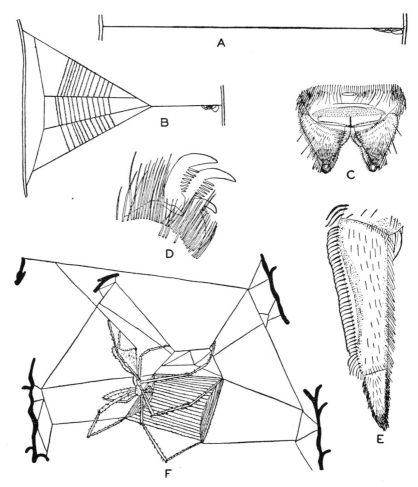

Fig. 6. The cribellate spiders.   A. Line trap of *Miagrammopes*.   B. Triangle web of *Hyptiotes*. C. Cribellum and spinnerets of *Amaurobius*.   D. Tarsal claws of *Amaurobius*.   E. Metatarsus and tarsus of *Hyptiotes*, showing calamistrum.   F. Retiarius snare of *Menneus* (after Akerman).

to come from the ampullate glands. To spin its composite hackled band, *Amaurobius* holds the hind leg of one side at right angles to the long axis of its body, with the tarsus resting against the metatarsus of the leg of the other side. The metatarsal comb is then rubbed back and forth over the cribellum, drawing out two ribbons that are attached to two lines of dry silk issuing at the same time from two other spinnerets. After a period of incessant spinning, the spider shifts to the other leg, supporting it as before by resting the tip across the opposing metatarsus. The result is a fairly regular, ribboned band of silk that appears as a single thread with a bluish color to the naked eye.

The cribellate spiders have retained more units of spinning equipment than have any other true spiders, and have maintained all of them as functional organs. It is thus not surprising that none have become truly vagrant, and that all rely to a large extent on their viscid threads to capture insects. The fact that they have retained the cribellum with its glands of sticky silk, indicates their reliance on it in some measure to entangle their prey.

In the cribellates, the unpaired claw on the tarsus is usually present, but it is reduced in size or even lacking in a few members that may be taking steps toward becoming vagrant types, or have learned to do without these tarsal hooks in their webs. Some cribellates are confirmed aerial spiders and spin tangled webs, sheet webs, and even orb webs, from which they hang upside down. Other cribellates run over an irregular blanket of webbing in an upright position. The web structures of the cribellates closely parallel those of the ecribellate spiders, the only difference being the incorporation of the hackled band by the former.

The cribellate spiders are quite sociable creatures. During the mating season the males enter the webs of the females and live there as partners until—presumably— they die a natural death. This tolerance carries even beyond the mating season inasmuch as among the cribellates we find nearly all our social spiders. Some live together in large colonies similar to those of certain gregarious caterpillars.

One of the controversial and perplexing problems in spider phylogeny has to do with the origin of the cribellate spiders and their relationship to the ecribellate families. By some they are held to be a homogeneous group derived from a single line of ancestral spiders that put their fading anterior median spinnerets to a new and original use by inventing the calamistrum. On the other hand, the cribellate spiders can be regarded as a remnant held over from a time when all spiders were cribellate; the modern forms then can conceivably originate from one or several distinct lines. The presence among these spiders of some in which the second pair of book lungs is still persistent suggests a very early origin for the group, and also strongly indicates that all ancestral true spiders were provided with cribellum and calamistrum. If we subscribe to this belief, and it is a very convincing one, then it can be put that the ecribellate spiders have *lost* the spinning organs, rather than that the cribellates have gained them. Some modern students are so committed to this former view that they place some of the cribellates in families with ecribellate ones.

THE FOUR-LUNGED TRUE SPIDERS

The American representatives of the family *Hypochilidae* are strange relicts of the past whose forebears were probably aerial contemporaries of the Paleozoic ground tarantulas. Four species of *Hypochilus* occur in the United States, and they have as their nearest relatives single species of the genera *Ectatosticta* and *Hickmania* found in China and Tasmania, respectively, and these are now assigned to separate families. Although we regard these relicts as being true spiders, they share many of the features of the mygalomorph spiders, the most notable being possession of the posterior pair of book lungs, which in all other true spiders have been transformed into tracheae. These lungs are situated beneath and about at the middle of the abdomen, and their spiracles open at the sides of a prominent furrow. The chelicerae are provided with venom glands entirely contained within the basal segment, and the heart has four pairs of ostia as in the tarantulas. Perhaps the most distinctive badge of the true spider is the articulation of the chelicerae. In *Hypochilus* we find the chelicerae largely intermediate in type between those of the true spiders and those of the tarantulas; since the fangs do not point toward each other, they are in many respects nearer those of the latter. The cribellum of this spider is a rounded plate lacking the median dividing ridge but is pinched before and behind to indicate the original dual character. It sits upon a low elevation that strongly suggests the segment of an ordinary spinneret.

Hypochilus shows a greater difference from the mygalomorph spiders in its habits of life than in its physical features. Whereas no tarantula has become a confirmed aerial cobweb spider, the hypochilids and a great many other true spiders have. It is quite possible that spiders resembling the hypochilids were the first to break away from the conservative tarantulas, and that they are modern representatives of an ancient group that gave rise to all true spiders.

Our best known species, *Hypochilus thorelli*, is found in the lush ravines or mountains of the southeastern United States, where it is quite abundant at elevations from 1,000 to about 5,000 feet, and especially so in the Great Smoky, the Nantahala, and the southern half of the Blue Ridge Mountains. It prefers shaded situations under overhanging rocks, and natural arches in forested areas near streams. Its webs—conspicuous objects even from a distance—are often found close together under the rock ledges. The webs are shaped like lamp shades with the top pressed against the overhanging surface, and consist of a very heavy mesh of cribellate threads over a base of dry silken lines. The spider hangs underneath this net with its long legs touching the sides of the aerial portion. In the great length of its banded legs, *Hypochilus* resembles most closely some true spiders of the family *Pholcidae*. Our species of *Hypochilus* do not seem to have the power of autotomy, and their legs do not break off at a point weakened and predestined for this purpose, as in other spiders. The males, which mature in the fall, differ little from their dull, yellowish mates. The male palpus is of very generalized design, and is provided with a conductor of the embolus as in the atypical tarantulas and most true spiders. The epigynum of the female is quite simple and presents no external openings; inside are four seminal receptacles—a feature shared by the atypical tarantulas.

Four species of *Hypochilus* are at present known from the United States. *Hypochilus thorelli* comes from the southern Appalachian region and has to the north of it in Virginia and West Virginia the similar species *gertschi*. A species of much the same appearance, *petrunkevitchi*, ranges in California from San Bernardino County northward in the Sierras to Humboldt County, and seems to favor the same kind of sheltered, mesic habitats as its eastern cousins. A quite distinctive species, *bonneti* of southern Colorado, lives in caves and sheltered rock wall surfaces and was found in good numbers at the bottom of the Black Canyon National Monument, at an altitude of 6,547 feet. The distinctive lamp shade web and the messy series of egg sacs hanging nearby quickly identify these remarkable spiders.

THE FILISTATIDS

One of the very common house spiders of our southern states is *Filistata hibernalis*, a large animal whose webs are often prominently outlined on the outside walls of buildings. This spider hides in a crevice usually at the center of the flat web during the day, and at night comes out to spin on its web, which soon gathers to its sticky lines an unkempt covering of dust and debris. The web, often more than a foot in diameter, is composed of a series of regular, radiating lines of dry silk over which has been spun many lines of cribellate bands. The touch of an insect vibrates the web and the disturbance is communicated to the hiding spider.

The hackled band of *Filistata* is composed of four different kinds of silk. The cribellum is combed with a very short calamistrum, and many tiny loops are produced, which, bundled together, give a most irregular shape to the characteristic threads. The spider lays down a dry line of two threads, retraces its steps upon this, and then puts down the irregular hackled lines, thus accomplishing its purpose in three operations rather than in a single one as does *Amaurobius*. The female of *hibernalis* is about one-half an inch long and is quite variable in size and color, varying from yellowish to brown or velvety black. These spiders live several years, increase gradually in size, and its youthful yellows gradually change to the darker colors of old age. Small, pale females, often found with egg sacs or emerging young, contrast strikingly with old blackish females with similar eggs or brood. The male is pale yellow, smaller, and his slender body is fitted with much longer legs, which, during courtship, are used to hold the front legs of the female while the couple parades back and forth in a prenuptial dance. These generalized spiders have simple genitalia: the suboval bulb of the male palpus is drawn out into a thin spinelike embolus; the epigynum consists of a pair of small receptacles with openings hidden under the genital lip.

Three distinct genera of the family *Filistatidae* are found in the United States. Several species of *Filistata* similar to *hibernalis* range over the western states. All of these are quite large spiders which most often are found under ground objects in the open and much less frequently around buildings. A couple of species of *Filistatoides*, small elongated species often with maculate abdomens, barely reach the southern fringe of our southwestern states. A third genus, *Filistatinella* with the species *crassipalpus* from the lower Rio Grande Valley of Texas and three other undescribed ones from the southwestern states, have some unusual features. These are small, dark, stocky species about one-third of an inch long. The tibia of the male palpus is greatly thickened and has at its base a curved spur into which the thick embolus, now turned backward, fits in its resting position. The webs of the species of *Filistatinella*, resembling those of the other genera except in size, are placed on rock walls often along streams and also on houses in humid situations.

THE TYPICAL HACKLED BAND WEAVERS

In this group are included the great majority of cribellate spiders and almost all of those that occur in the temperate regions. About 300 species are known from the United States, but for the most part these represent only a few different types. The prevalence of their meshed webs on the ground and on vegetation everywhere is an index of their abundance and comparative success. With few exceptions, they are confirmed web spiders and stay in their snares most of the time, walking upright over the bluish sheet. Their sedentary bent has not molded them into such aerial types as the uloborids, but some do climb vertical meshes and are at home in aerial tangles. Three claws are almost invariably present on the tarsi, but they are not aided in their climbing by accessory claws. Many are swift runners that can be seen dashing across paths with the celerity of vagrants; most of these are males. The habits of these hackled band weavers are of considerable interest but they are not known to do things quite as amazing as do some of the aerial web spinners.

Four different families occur in our American fauna, the *Amaurobiidae*, *Tengellidae*, *Dictynidae*, and *Oecobiidae*. Before considering these families, mention will be made of two unusual families not found within our territory. The first of these, the *Eresidae*, is made up of robust, moderate to large spiders, similar to jumping spiders, and often quite brightly colored, especially the males. Some of them are fine tunnelers and are even known to use a trap door to close their tubes. Many spin sheet webs connected with tubular retreats. Some species of *Stegodyphus* join together and spin an immense communal web over bushes, forming an irregular saccular retreat partitioned in various ways, in which many individuals live amicably together. The second exotic family is the *Psechridae* with several large, conspicuous representatives common in the Pacific region. These spiders often spin a huge web, at the center of which is a flat sheet similar to those made by the sheet-weaving *Linyphiidae*. The spiders creep over the ventral surface of the sheet, hanging back-downward, as do the aerial ecribellate spiders, on greatly elongated legs of which the terminal segments are flexible. One of the strange features of the pschrids is the presence of well-developed claw tufts on the tarsi. These are probably used for a very different purpose than are the claw tufts of the wandering spiders. Perhaps they aid in unfastening the median claw, and serve in the same way as the accessory claws of the orb weavers.

The largest typical cribellate weavers of the family *Amaurobiidae* belong to the genus *Callobius*. Many of the females are robust creatures attaining a length of three-fourths of an inch. Their colors are usually browns or blacks, but the abdomens are variegated with a dorsal series of yellowish chevrons forming a pale band. The males, not much inferior in size, are usually in evidence only during the fall and early spring, and then are rarely seen during the rest of the year. A single species, *Callobius bennetti*, is common over our eastern states; at least twenty-four additional ones abound in the mountains of the western states and in the northern woods. The family is strongly represented in North America with as many as fifty-two additional species assigned to such genera as *Amaurobius, Callioplus, Pimus, Titanoeca*, and *Zanomys*. Most of these come from the Californian region.

Callobius bennetti spins a large irregular web in dark, moist situations. Whereas much of the silk may be hidden from sight, not infrequently the web is placed in plain view against a vertical surface. The dry framework of the snare is commonly put down as a series of lines from the central retreat, in which the spider stays most of the time. Over the dry lines the cribellate threads are spun loosely, making a thick mat on which the spider runs. The spinning activities are best observed at night; then the carding can easily be seen in the rays of a headlamp. At the same time males may be found near the web of the female. The egg sac is a flattened bag, attached to a stone and usually covered over with a mesh of threads. The females stay with the eggs for long periods, often being found with the sac some time after the young have hatched.

Some amaurobiids, brought to us by trade from widely separated parts of the world, now seem to be established residents of this country. As early as 1871 a large domestic species, *Amaurobius ferox*, was reported from Providence, Rhode Island, and it is now common in our northeastern states. Derived from Australia or New Zealand was *Ixeuticus martius* now widespread in coastal California and northern Baja California. More recently, a species from Argentina or an adjacent country, *Metaltella simoni*, has been found in the Gulf states from coastal Florida to Louisiana and Mississippi, where they spin small webs under logs. Many species are constantly being brought into this country by trade but few of these find the climate to their liking.

Australia is particularly well supplied with spiders related to *Amaurobius;* the habits of certain varieties are of special interest. In the Jenolan Caves of New South Wales lives a gregarious species, *Amaurobius socialis*, which spins great webs on the roof. These giant reticules, one of which measured twenty feet in length and more than four feet at its greatest width, hang from the roof and are draped over the stalactites. They are closely and densely woven to the consistency of a heavy fabric, such as a shawl, and are filled with openings through which the spiders retreat to the interior. Mating, egg-laying, and emergence of the young all occur within the limits of the web, as among many truly social spiders. Also, in New South Wales may be found gregarious species, perhaps relatives of *Amaurobius*, that infest orange groves. They mat the limbs with a thick covering of web so densely woven that it affects the normal respiration and development of the tree. The leaves wither and fall, leaving the branches dead and covered with unsightly webbing. Such spiders can become a pest of nearly equal rank with tent-building insects.

The family *Dictynidae* is represented in the United States by at least 150 species, all of which are quite small spiders, perhaps averaging about one-eighth of an inch in length. Some few of these are brightly colored in reds, tans, and browns, with here and there a brilliant yellow spot or band, but for the most part their bodies are dull, clad plainly in gray hairs. Several different groups of these small mesh spinners are known but they are all similar in appearance and habits. Some of the smallest live under debris on the ground, where they spin tiny webs and are rarely noticed; others of larger size spin on the walls of buildings and on plants. Their lacy meshes are conspicuous objects but, aging, they become obscured with dust and debris. Mention can be made here only of a few species whose habits are illustrative of the whole group.

Dictyna annulipes, a small species with a large oval abdomen, has its dull body quite completely masked by a covering of light gray hairs, which on the carapace form three distinct stripes, and on the abdomen outline a pattern of darker chevrons. Favorite sites for its web are board fences and the walls of buildings. A tiny crevice between boards will provide this spider with an adequate retreat from which it can lay out the dry foundation lines of its snare. These lines frequently radiate with the most precise regularity and, when crossed evenly with the thick hackled bands, reveal a web as delicately spun as a lace doily. This little spider often chooses the outside of a window sash as a location for its snare, and lashes it to the smooth glass as well as to the adjacent wood.

Early summer is usually the pairing season for these friendly spiders, and at this season the male may be seen in the web with the female. He resembles her closely in general appearance, but is somewhat more slender and has longer legs. His head is often quite elevated and arched over the long, curved chelicerae. These latter are provided with a stout spur near the base; they turn upward at the tips and curve strongly outward at the middle, leaving a conspicuous round opening between. In some species of this particular group of *Dictyna*, the modified chelicerae are known to be used to hold the chelicerae of the female during the mating, and it may be presumed that they render the same service for our own species.

A close relative of *annulipes* is *Dictyna volucripes*, a slightly larger spider, similarly clothed in pleasing gray raiment, which prefers open sunny fields for its home. The usual sites are the ends of dried weeds (Plate XVIII) and grasses, and especially the dried stems left over from the previous growing season. Upon this harvest skeleton *volucripes* spins its characteristic mesh; the foundation lines bridge from stem to stem and over them is woven a crisscross of viscid bands, to form perfect little lattices and other pleasing symmetries. During the summer white, lens-shaped egg sacs are hung in the deeper part of the tangle, and after the young hatch they spend some time in the web with the mother.

Much more brightly colored than either of the above species is *Dictyna sublata*, often light brown in color with a yellow marked abdomen and white legs. This spider hides its web in the leaves of bushes instead of placing it in the sun. It will find a leaf with slightly rolled edges, then spin a sheetlike web across the opening to form a shallow bowl; in this it remains and here its eggs are placed.

The villagers of certain mountainous portions of Michoacan, Mexico, are plagued during the rainy season by immense swarms of flies that invade their homes. Their defense against these pests is unique. They rely upon the *mosquero*, a tiny cribellate spider (*Mallos gregalis*) one-sixth of an inch long, which lives in vast colonies on the twisted oaks and scrub trees at altitudes of about 8,000 feet. The nest of a *mosquero* community is often more than six feet square, and thickly invests branches or even a whole tree with a spongy inner layer of dry silken lines and an outer envelope of sticky hackled band threads. The villagers cut a branch from a tree, and suspend the animated fly trap from the ceilings of their homes. The accommodating houseflies alight on the sticky threads, whereupon they are enveloped and dragged into the inner galleries to become the prey of the colony. After the fly season is over and the spiders have become mature, the adults desert the colonial web, perhaps to start new colonies elsewhere. Their eggs and young remain, develop in the inherited nest, and are on hand during the next fly season. In the field the webs of the *mosquero* resemble those of processionary caterpillars. These social spiders also occur in Jalisco in the countryside of Guadalajara, where Wesley Burgess found one huge web covering the limbs and branches of three-quarters of a sixty foot tree. In the inner recesses of such communal webs live many small beetles of the genus *Melanopthalma* that are said to attend to the cleanliness of the nest by keeping it free of debris. These presumed commensals live on the small bits of food discarded by the *mosquero*. Also living in complete harmony with the colony in

Michoacan was one of the running spiders, *Poecilochroa convictrix*, which was supposed to be a commensal but its exact status is less certain.

The cribellate spiders of the family *Tengellidae* have brushes of scopular hairs beneath the metatarsi and tarsi of the legs and have the median claws of the tarsi reduced in size or even absent. These features would seem to indicate that they are hunters. Indeed, they are often compared to the hunting clubionids, which they resemble in general appearance and in superficial features. However, *Zorocrates*, represented by several species in our southwestern states, still relies on an expansive web to snare its prey. The web, resembling to some extent that of *Amaurobius*, is usually placed beneath stones, and made into an effective trap by spinning many viscid hackled bands over its dry framework. It may be that *Zorocrates'* scopular brushes contribute to better movement over the surface of the web, or perhaps that this spider is on its way to becoming a vagrant type and therefore spends part of its time outside the limits of its snare.

The related family *Acanthoctenidae* is known only from tropical America. In *Acanthoctenus* true tarsal claw tufts are present in addition to the thick scopular brushes beneath the metatarsi and tibiae. These flattened creatures often sit under bark, closely appressed to the surface, and move with great speed when they are touched, their claw tufts aiding them in holding on to surfaces, as with the ecribellate vagrants. *Acanthoctenus* combines a sedentary aptitude with running ability. It spins a loose web, embellished with sticky bands to entangle its prey.

At the end of the typical hackled band spinners we have the family *Oecobiidae*, a distinctive group of small spiders not fully one-eighth of an inch long, that have little in common with the other families. Several of the species live in habitations and have been transplanted by male to widely separated parts of the world by trade. Our best known species, *Oecobius annulipes*, is now widespread over the world in subtropical and tropical areas; in the United States it is found in our southern fringe of states and is especially widespread in California. It is a whitish or pale brown spider with distinct black spots on the abdomen. The broader than long, suboval carapace has the eight large eyes in a group near the center. The abdomen bears below near the end a remarkable, two-jointed, movable anal tubercle which bears a fringe of stiff hairs. Recently, it has been discovered that this fringe is used to comb the silk from the large posterior spinnerets much in the fashion that other cribellates use the calamistrum. This spinning invention supplements that of the cribellum, which is said to be called upon only rarely to enswathe prey. The microscopic webs of *Oecobius* are frequently spun over cracks in the sides of buildings; they are only about the size of a postage stamp, but seem quite adequate to entangle ants, flies, and other tiny insects used for food. The generic name *Oecobius* means "living at home," and well characterizes these dwarfs so often found in and on the walls of dwelling places.

More than a dozen species of *Oecobius* are now known from North America and several of these have recently been described from Mexico. The most interesting is *Oecobius civitas*, a tiny blackish species that weaves large webs beneath rocks on the shores of Lake Sayula, in Jalisco, Mexico, and in adjacent states. In such webs live from 60 to 100 individuals, including many subadult individuals and a few mature ones. The communal web comprises a large series of tiny webs, each the hiding place of a single individual. Although these spiders are gregarious and live close together in harmony, each is a solitary individual that catches its own prey and does not much enter into cooperative actions with its neighbors. A favorite food of these spiders is ants which are captured by means of sticky silk ropes combed out by the fringe on the anal tubercle.

THE AERIAL HACKLED BAND WEAVERS

The most pronouncedly aerial of all the cribellate spiders are those of the families *Dinopidae* and *Uloboridae*, groups largely tropical in distribution that press northward in small numbers into the temperate zone. Within the limits of the United States are found representative species, many of which are remarkable for their physical appearance and strikingly resemble bits of dried leaves, twigs, thorns, buds, scales, and similar natural objects. The common name of "stick spider" has been applied to some of them; the name characterizes the whole group rather well, even though not all are elongate. All do hang downward from a more or less intricate web, and in their movements on silken lines parallel closely such aerial ecribellates as the orb weavers and tangled web weavers. By some students these spiders are regarded as closely allied to their ecribellate cousins, but to others the extensive use of the hackled band sets them apart and indicates only a distant relationship. In all the groups the third tarsal claw is present, modified into a hook to suit the needs of confirmed sedentary types.

Some very ingenious methods have been developed for capturing prey. The dinopids have perfected a method of expanding and hurling a sticky net over flying insects. Not to be outdone by their cousins, the uloborids construct a splendid orb web rivaling in excellence that of the orb weavers of the family *Araneidae*. The triangle spider, *Hyptiotes*, has abandoned all the orb save a single sector of four rays. Even more niggardly is the tropical stick spider, *Miagrammopes*, which employs a single line on which it spins a band of sticky silk.

The Dinopids. The name "ogre-faced spider" has been applied to species of *Dinopis* (from the Greek, meaning "terrible appearance") because of their weird appearance and the enormous size of their posterior median eyes which, projecting forward like great headlights, render inconspicuous the remaining six eyes. The names dinopid stick spider or simply dinopid are preferred. Their nocturnal habits would seem to demand good night vision as furnished by such large eyes, but there is no positive evidence that this is true. Our *Dinopis spinosus* has been considered quite rare, but recent collecting has shown that it is not uncommon in our deep southeastern states and especially Florida. The dinopids are known by many species from tropical regions from all around the world; many have grotesquely formed heads and humped and lobed abdomens.

Our single species, *Dinopis spinosus*, has been studied intensively by Bert Theuer, who has left us a wealth of natural history and distribution information in an unpublished thesis at the University of Florida in Gainesville. *Dinopis spinosus* is a slender spider, which hangs head downward from a platform of dry silk on long, stiltlike legs. When mature, the female (Plate 8) is about two-thirds of an inch long, and frequently has the abdomen armed above and near the middle with short projections. The male is smaller and more slender than his mate, and his thin legs are at least three times as long as his whole body. A dark band running the length of the abdomen below, and a few lines and spots above, are the only distinctive pattern on the otherwise drably marked bodies of these spiders. During daylight hours the phlegmatic dinopids are usually to be found pressed flat against a branch or hanging from framework lines in their chosen casting site. The resemblance of these spiders to twigs, buds, spines, or other natural irregularities in the habitat must pay dividends by giving them some immunity from predators.

Completely quiescent during the day, *Dinopis spinosus* rouses to action at sundown, moves over the small tangle of dry silken lines, and prepares its capturing web. As described by Theuer, a favorite webbing site is an area about twelve inches square on the ivy-covered walls of buildings at the University of Florida. Moving upward the spider spins an inverted Y-shaped series of dry lines with three principal lines on each side. Then across the inner portions of four of these lines and still attached to them it spins transverse cribellar lines, thick bands carded from the cribellum by means of the comblike calamistrum. Thus is formed a sticky patch of webbing (Plate 8) about the size of a postage stamp: this *Dinopis* then picks up with her four front legs but this is still attached to the four dry lines. The spider turns and hangs head downward, secured above by hooks in her hind tarsal claws attached to dry lines, and thus hangs loosely with the capturing web held in the tarsi of the four front legs. Now positioned, *Dinopis* waits for the appearance of flying insects.

When a flying insect comes near enough, *Dinopis spinosus* throws its front legs far above the head, thus enlarging the viscid snare to several times its original size, and into it flies the prey. The snare is not thrown over the insect but simply held in a favorable position to intercept the victim. It might be conjectured that unknown pheromones, invisible chemical attractors, sometimes attributed to the success of *Mastophora* and her viscid globule, may here be called into action but at present there is no evidence for them. In any case, the victim becomes entangled and may be further aided by twitching of the snare by the spider. Small prey such as mosquitoes and other dipterons are plucked out of the web and transfered to the chelicerae for eating. For large and vigorous captives, such as moths, the spider maneuvers its legs to further enmesh the victim, at this time cutting the guy lines still holding the corners of the snare, and spinning other dry line to hold the insect or in extreme cases wrapping the viscid snare around the victim. The bundled prey may be bitten to quiet it or simply flooded with digestive juices to start the eating process. If the viscid snare has not been too much damaged or used for wrapping the victim, it can be used again and again. A much damaged snare is bundled up into a small wad and eaten, or it may be dropped to the ground. At daylight the spider ceases its capturing techniques and rests hanging head downward on its webbing site.

The dinopid method of snaring a victim (Fig. 6, F) was compared by Akerman to

enveloping it as the Retiarius with his net enveloped his opponent before piercing with his trident in the Roman gladiatorial combats, or better, like the old-fashioned butterfly net on two sticks, held by the two hands, which was thrown over an insect to catch it.

Although having no exact counterpart among spiders of any other family, this device parallels in a general way that of the bolas spiders (*vide infra*).

Hackled Band Orb Weavers. An outstanding achievement of the *Uloboridae* has been the invention of an aerial orb web equal in symmetrical beauty and similar in fabrication to that of the ecribellate orb weavers. It is believed by many that this creation was a novelty separately arrived at, not one result of ancient habits common to allied spiders that later diverged. The germ of the orb web is observable in the great regularity of the webs of hackled band spinners less versatile than the *Uloboridae*. Even the irregular mats of *Amaurobius* and *Filistata* are based on a framework of dry rays arising from a central retreat. Many species of *Dictyna* spin aerial sheets of such regularity that in form they approach sectors from the webs of the uloborids. Once a symmetrical design had been realized, lifting the orb web from a surface to an aerial station was a relatively simple step. The orb web of *Uloborus* most often lies horizontal, or slightly inclined, and is only rarely the vertical structure of the typical orb weavers. The horizontal position is a less favorable one, dependent for success on insects that fly upward against it or drop down upon it; whereas the vertical web can intercept a much larger flying fauna.

Featherfoot Spiders. As has been noted, the curious spiders of the genus *Uloborus* are most numerous in the tropics of the world, relatively few varieties occurring in the north. Several distinctive species are found in the southern United States, but only one, *Uloborus glomosus* (formerly known as *americanus*), rather appropriately named the "featherfoot spider," is common all over the eastern United States and southern Canada. This uloborid has a carapace longer than broad, which is provided in front with eight eyes, in two rows, whose small size confirms the slight reliance this aerial creature places on eyesight. Its chelicerae are moderately robust, but no venom glands are associated with them—a condition almost unknown in other spiders, and suggesting that the sticky spiral of the orb web and the chelicerae alone are adequate to quiet its prey. The long front legs are often curved, in many species being provided with the tufts of feathery hairs that are the source of the common name. The abdomen is often surmounted with humps and bedecked with pencils of hairs. A pronounced variation in coloration is characteristic, and pale white, speckled, lined, or all black specimens are often found in the same species. A congener is western *Uloborus diversus* (Plate XXI).

The relatively small orb webs of the featherfoot spider, four to six inches in diameter, are usually placed close to the ground in moist, shaded situations—on low bushes and underbrush, on dead sticks, in hollow stumps, or among rocks. The invariably horizontal web is composed of the same elements as that of the typical orb weavers: foundation lines, radii, dry spiral scaffolding, and a concentric series of sticky spirals. Before laying down these latter, *Uloborus* spins a typical preliminary spiral scaffold of dry silk that is used as a bridge to the next radius. The spirals are a composite of viscid and dry threads, like those of the *Araneidae*, but here the sticky material is carded from the cribellum with the aid of the very regular row of calamistral hairs on the fourth metatarsus. The spider has never depended on artistic perfection for its capturing snare, but rather on the sticky lines, and close examination shows that the web is imperfect in many respects.   Quite often the web is most unsymmetrical—especially during the cocooning season, when the feeding instinct is replaced by a maternal one—and irregular in its details, but it remains quite as pleasing to the eye as the snares of the typical orb weavers.

Special features of the web are the hub, which is closely and beautifully meshed, and the ribboned decorations or stabilimenta that ornament the web and possibly add to its strength. The most frequent form of the stabilimentum is a scalloped band that crosses the central portion of the orb; it is scarcely visible at the delicate hub. Other variations are numerous, a common one being a ribbon coming from a nearby sector to form a V-shaped figure; or four ribbons forming a cross; or broken or completed circles around the hub.

In position, the featherfoot spider lies upside down, stretched out horizontally beneath the hub of her web, her legs directed forward and backward to form a bridge between the stabilimenta and make a complete band across the snare. As she hangs there, swaying with the breeze, she often resembles a small stick. When her eggs are laid, she places the several elongate sacs in a row across the web, and then aligns her long body so that she becomes almost indistinguishable from them—one in a line of bits of debris.

The uloborids are friendly spiders so it is not surprising that individuals live close together and that some live as social spiders in large aggregations. In the foothills and mountains of southern Arizona and adjacent Mexico *Uloborus oweni* invests herbs, shrubs, and cacti with a loose communal web in which juvenile to mature stages live close together. The females hang their egg sacs, consisting of two to six or more bags closely knit together into a long parcel, from dry lines of silk. The small inclined orb webs seem to be the property of single individuals. A somewhat larger, whitish species, *arizonicus*, often lives in large aggregations under culverts or in natural depressions. Some tropical uloborids of the *republicanus* group, to which our Arizona species belong, spin immense webs in which large numbers of males and females live amicably together. These colonial webs feature a large central retreat suspended from many long silk lines running in all directions and forming a loose maze. Most of the males, as well as many females and spiderlings, live in the inner part of the web; but from time to time—and this is a particularly fascinating part of their activity—individuals detach themselves and move to open spaces on the periphery, there to spin their own characteristic round webs. The outer part of the communal web provides snare space for all the spiders, and part of the time they live snugly in their tiny orbs. Mating takes place in the central retreat, and the egg-laying occurs there as well.

Triangle Spiders. Almost anywhere in the United States may be found the triangular snare (Fig. 6, B) of the spider *Hyptiotes*, which because of its small size and retiring habits, is far less familiar than the web it spins. Rarely more than one-sixth of an inch in length, the triangle spider hangs back-downward from a dried twig in its favorite trapping site, and is no more noticeable than a bit of dead wood, a bug, or a piece of bark. The carapace is broad and low; it supports a thick, oval abdomen on which are usually visible slight humps set with a few stiff hairs. Drably clothed in grays and browns, *Hyptiotes* harmonizes rather well with the dry branches of its home, and affords a striking illustration of close resemblance to environment. All eight eyes are present in this species, but one minute pair is so well hidden in the hair covering that the spider was once thought to have only six eyes. The male ordinarily becomes adult in the early fall, and at that season may sometimes be found near the web of the female, which sex he resembles closely except for smaller size.

Four species of *Hyptiotes* occur in the United States and Canada. The common species in our eastern states is *Hyptiotes cavatus*, the triangle spider of Hentz, which lives in deciduous trees and underbrush and in pinewood habitats, often favoring dried stems and branches. A boreal triangle spider, *gertschi*, is abundant in the western part of the country, and largely replaces the other species in eastern Canada, where it occurs as far south as Maine and New York. It lives on dead branches of pines and other conifers in southern Ontario, and in the western United States it uses these habitats as well as suitable stations on cliffs, in ravines, and under bridges.

The web (Fig. 6, B) of the eastern triangle spider, *Hyptiotes cavatus*, is best understood by comparing it, as did Professor Bert G. Wilder, to an ordinary pie. The orb of *Uloborus* is an entire pie; that of *Zygiella*, one of our typical orb weavers, is a pie with a piece cut out of it; and that of *Hyptiotes* is the missing piece. This triangular web consists of a fifty-to-sixty-degree sector with radii twelve to twenty inches long. It invariably consists of four rays of dry silk, across which are laid down ten or more viscid lines of hackled band that correspond to sections from the spiral line of an orb web. The four rays are attached to an arc line tied to twigs, and converge near a point on a single bridge line fastened to some nearby object.

The spinning of the web, often accomplished during the early hours of the evening, is a most interesting process, and the details corroborate the belief that its structure is derived from the uloborid orb web. The first line is a bridge from the resting site of the spider to an adjacent dried twig. It is customary for *Hyptiotes* to place the bridge line by hand, moving around the periphery of her hunting grounds to the point of attachment and then pulling the line tight; but in many instances air currents are called upon to balloon the line to a mooring point, as is the practice of the typical orb weavers. A vertical thread from one end of the bridge line is tied to a twig, and forms the arc of the sector. The third principal line returns to near the other end of the bridge line and completes the triangle. Then, between the radial lines, *Hyptiotes* places two more rays.

At this stage the spider has constructed a sector of four rays attached at one end to a single line and at the other to an arc thread. Upon this framework must now be placed the viscid threads that will make the web a trap. But before the sticky lines are added, *Hyptiotes* spins a row of three or four dry scaffolding threads, extending from the apex toward the middle of the triangle, that serve to steady the web by holding the radii in place, and that will simplify the laying down of the cribellar silk by providing a bridge from ray to ray. These scaffolding threads are analogous, or even homologous, to the dry spirals or spiral bridge of the ecribellate orb weavers, and are put down in the same sequence from the apex of the triangle (or hub) outward, and are eliminated in much the same way—bitten out when the web is finished.

To lay down the viscid sections, *Hyptiotes* crawls along the uppermost ray nearly to the point at which it joins the arc line, spins and attaches a band of sticky silk, then crawls back toward the middle of the triangle, spinning as she goes and holding the thread free of the ray. When she reaches the outermost scaffolding thread, she descends upon it to the ray immediately below, and upon this returns, reeling the sticky line back in, until she is immediately below the first point of attachment. Here she fixes her line. In order to put down this first vertical cribellar thread, which extends only three inches or so between the two upper rays, *Hyptiotes* must often crawl forward and backward a dozen inches. One might ask at this point why *Hyptiotes* does not drop down directly to the ray below? The triangle spider knows her web only by the touch of the silk, cannot see or know the position of the other rays, and is dominated by instinctive actions that keep her pursuing a prescribed course.

The spider continues this roundabout process until the four rays are bound together by a slightly zigzag vertical line of three sections. She then crawls around the triangle to the top ray once more, and starts on a second line, using her legs to measure its distance from the first. The final number of these partial spirals varies from ten to more than twenty. As the series meets the scaffolding, these latter threads are cut out of the web.

The finished web is an extraordinary structure, and is employed in a remarkable way to provide the spider's food. *Hyptiotes cavatus* takes up a position at the end of the bridge line, near the apex of the triangle, her hind legs touching the silken anchor almost in contact with the twig. With her front legs she pulls the line until the whole web becomes taut; then, holding the slack thus gained over her body, she settles down to wait for her prey. A small moth or other flying insect strikes the web and adheres, struggling violently in the viscid coils. Immediately *Hyptiotes* lets go of the slack. The web snaps forward, carrying the spider out a short distance with it, and the vibration of the sticky swaying line causes the victim to become more firmly enmeshed. *Hyptiotes* seems able to estimate the character of the insect from the nature of its frantic struggles and acts accordingly. The snare may be drawn tight once more and snapped, and this action will be repeated again and again until the satisfied spider is ready to crawl over her lines to the victim.

Hyptiotes never bites her prey as do many other web spiders, a fact undoubtedly related to the absence of poison glands in this family. Instead, she comes up to the insect, turns her back on it, and rolling it over and over with her legs, covers it with a thick bluish web. Completely helpless, the victim is carried back to the resting site and sucked dry in the leisurely manner characteristic of the triangle spider. This method of overpowering prey by means of thick bands of silk is analogous to the habits of the comb-footed spiders and the typical orb weavers. Sometimes *Hyptiotes* will bundle up the web and hurl sections of it over a struggling victim, thus destroying the web and making necessary the spinning of a new one for her next period of trapping. Inasmuch as one victim provides food for one or several days, the loss of the snare does not much handicap her.

One wonders whether *Hyptiotes* has not gone to more trouble than the web is worth in producing its triangle trap. Although it will probably catch more insects because of its vertical position and greater size, almost as much spinning and silk go into its fabrication as is expended in the horizontal web of *Uloborus*. Furthermore, the trick of snapping the trap in order to further enmesh the prey may well be an unnecessary precaution; and the careful enshrouding of the bound victim is likewise an act of doubtful necessity. It is true, however, that the same kind of objection to needless efficiency can be leveled against the snare of the typical orb weavers.

The Single-Line Snare. The stick spiders of the genus *Miagrammopes* are creatures of the tropics, and although they are common enough in the West Indies and in Mexico, they barely reach the United States in the lower Rio Grande Valley of Texas. They have produced a marvelous trapping device that represents even a greater simplification of the orb web than does the snare of *Hyptiotes*. The four-rayed triangle is reduced to a single line.

The stick spiders resemble *Hyptiotes* in general structure, but they are more elongate, and are thinly covered with dull grayish hairs over a dusky brown body, so that they almost perfectly resemble small, thin sticks. On the carapace are four pairs of eyes, the two front pairs being so small and so well hidden that only the hind ones are easy to discern. The front legs are long and thick, stretched forward in close contact with each other; against them press the short second pair; while the hind legs extend backward along the sides of the abdomen, and fit closely against the body to enhance the remarkable sticklike appearance.

The snare of *Miagrammopes mexicanus* (Fig. 6, A) probably typifies that of the whole genus. It is a single horizontal line, attached at both ends to twigs, and stretches about four feet across open spaces in the forest. Across the center of this horizontal line for eighteen inches or more the spider cards out a heavy band of viscid silk. The next step resembles the triangle spider's method: *Miagrammopes* moves to the end of her foundation line, and, assuming a position in which she almost touches the mooring twig with her hind legs, appears to be a continuation of it. She draws the line very taut, until she has a loop of slack to hold over her body. The thick center of the snare offers an attractive and familiar looking resting place for gnats, flies, and a whole host of flying insects. Whenever one alights, the stick spider lets go the loose thread and shoots forward with the elastic line for about one-half an inch. The released tension jerks and sways the thread, causing the victim to become more completely entangled. *Miagrammopes* then rushes to the site of capture, and again like *Hyptiotes*, further enmeshes the unlucky insect in bluish silk, which she reels and combs from her spinning organs. When her victim is completely helpless, she cuts it loose and holds it in her chelicerae and palpi. She then closes the rent in the trap, and crawls back to her retreat, where she adjusts the line for the next capture.

9
The Aerial Web Spinners

The vast assemblage of spiders treated in this chapter are those that spin silk from their bodies and produce many types of aerial webs. Whereas their relatives developed alertness, speed, brute strength, and a minimum use of silk to become hunters, the sedentary types on their gossamer lines swung far aside from that line of ecribellate spider evolution. Theirs is a story of silk: on tiny claws that have become increasingly effective as hooks they hang upside down from the threads of a circumscribed web, rarely leaving its confines voluntarily. Their sense of sight is rather poor. For this deficiency they have compensated by spinning expansive tangles, sheets, and formal web designs to enlarge their area of action; the struggles of an insect in the farthest recesses of the snare are communicated to them. Within the confines of the web the sedentary spiders have become supreme autocrats.

They are a motley crew running to all sizes and shapes. Many are shy, lie immobile in the web, and when disturbed drop on dragline threads to the security of deep underbrush. Others stay hidden away in retreats or under objects until their traps are touched by small animals. Some are quite agile and run nearly as well as hunting spiders; others, in appearance well proportioned for running, have legs too long and thin. A few specialize in inaction; they hang like inanimate slivers or clods in their webs, to all intents part of the debris that adheres to the lines. Others, possessing greatly elongated abdomens that they wave gently back and forth, resemble in form and action common caterpillars. Most are fat creatures with short legs that seem modeled for an acrobatic life. The great majority are tiny and inoffensive; therefore, they rarely come to our notice; but some of the orb weavers are giants, in bulk exceeding some large hunters.

The webs of the sedentary spiders, displayed on every side in myriad sizes and designs, vary from crude artistry to extraordinary workmanship. Such diverse structures did not come into being at a single stroke; they are the results of long,

Plate I

Lee Passmore

a. A desert solpugid, *Eremobates*

Lee Passmore

b. A tailed whip scorpion, *Mastigoproctus giganteus*

RELATIVES OF SPIDERS

Plate II

a. Molting. Carapace and chelicerae freed

b. Molting. The shed skin   c. Cradle of eggs in burrow

TRAP-DOOR SPIDER, *Bothriocyrtum californicum*

Plate III

a. Exposed burrow

b. Male

c. Cork-door nest held open

CALIFORNIA TRAP-DOOR SPIDER, *Bothriocyrtum californicum*

Plate IV

a. Capturing a ground beetle

b. Lifting the cork lid

CALIFORNIA TRAP-DOOR SPIDER, *Bothriocyrtum californicum*

Plate V

A Western Trap-Door Spider, *Aptostichus*. Dorsal View of Male

George M. Bradt   *George M. Bradt*

a. Surprised in its burrow   b. Exposed burrow

A MEXICAN TRAP-DOOR SPIDER, *Eucteniza*

Plate VI

H. K. Wallace

a. Trap door spider, *Myrmekiaphila*

b. Mexican tarantula, *Aphonopelma*

MYGALOMORPH SPIDERS

Plate VII

Dick Freeman

a. Clambering over stone

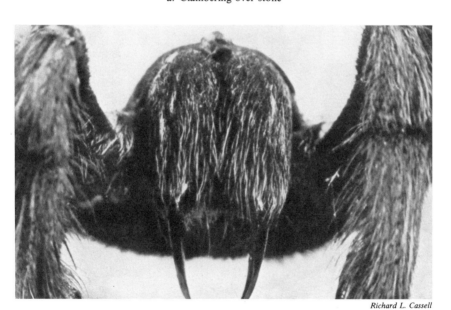

Richard L. Cassell

b. Portrait

MALE TARANTULA, *Aphonopelma*

Plate VIII

George M. Bradt

Lee Passmore

a. Female on desert soil   b. Web-covered entrance to burrow

Lee Passmore

c. Female and egg sac in exposed burrow

TARANTULA, *Aphonopelma*

Plate IX

Lee Passmore

a. The tarantula assumes a defensive attitude

Lee Passmore

b. The wasp inserts its sting

Lee Passmore

c. Pulling the bulky prey to prepared burrow

TARANTULA, *Aphonopelma,* AND TARANTULA HAWK

Plate X

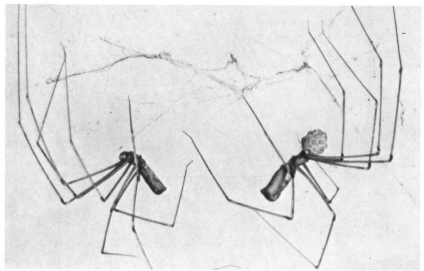

a. Male and female, with eggs, in tangled web

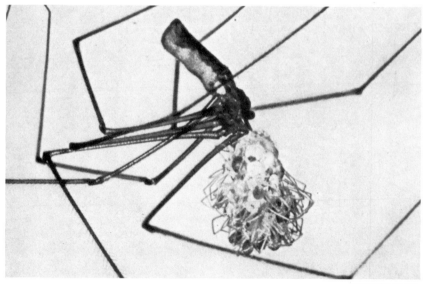

b. Female holding mass of recently hatched young

LONG—LEGGED CELLAR SPIDERS, *Pholcus phalangioides*

Plate XI

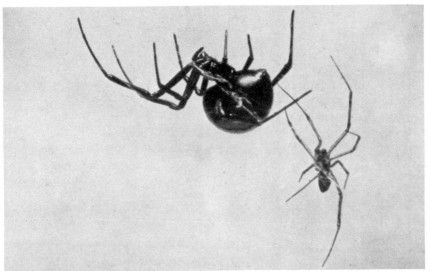

Walker Van Riper

a. The cautious approach of the small male

Walker Van Riper

b. The mating

COURTSHIP AND MATING IN THE WESTERN BLACK WIDOW, *Latrodectus hesperus*

Plate XII

Walker Van Riper

a. The male after mating, as here, killed and eaten by the female

Walker Van Riper

b. A female in her tangled snare with long-legged spiders, *Psilochorus*

WESTERN BLACK WIDOWS, *Latrodectus hesperus*

Plate XIII

a. The spider approaches as the cricket touches the capture threads

b. Nooses of swathing film are combed over the leg

A COMB-FOOTED SPIDER, THE WESTERN BLACK WIDOW, *Latrodectus hesperus*, CAPTURES A JERUSALEM CRICKET

Plate XIV

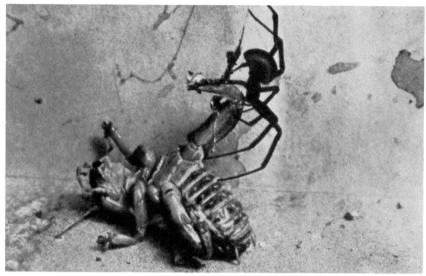

c. Tiny fangs inject the venom

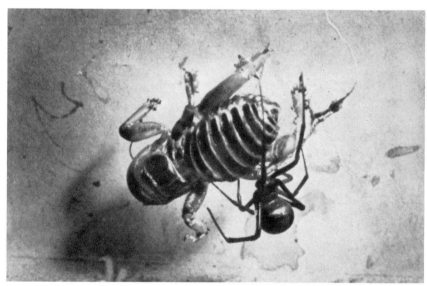

d. The bulky insect is lifted above the floor

A COMB-FOOTED SPIDER, THE WESTERN BLACK WIDOW, *Latrodectus mactans*, CAPTURES A JERUSALEM CRICKET

Plate XV

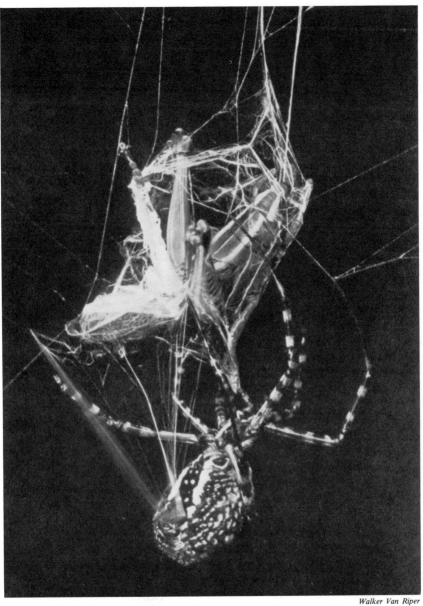

Walker Van Riper

BANDED ARGIOPE, *Argiope trifasciata,* SWATHING A GRASSHOPPER

Plate XVI

Edwin Way Teale

ORANGE ARGIOPE, *Argiope aurantia,* WITH SWATHED PREY

Plate XVII

H. H. Harrison

CLUSTER OF BABY ORB WEAVERS, *Araneus*, PREPARING TO DISPERSE

Plate XVIII

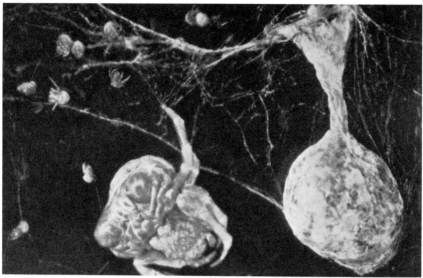

Female bolas spider, *Mastophora cornigera*, with recently emerged brood, including some adult males

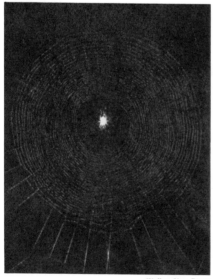

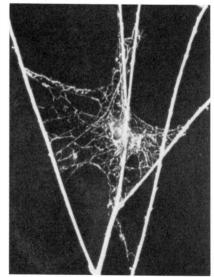

A symmetrical orb web of a mountain orb weaver, *Aculepeira aculeata*

Meshed web of *Dictyna* on dried weed

ORB WEAVER SPIDERS AND TANGLED WEB

Plate XIX

Lee Passmore   *Lee Passmore*

a. Female and pygmy male   b. Egg sac
SILVER ARGIOPE, *Argiope argentata*

Lee Passmore   *Lee Passmore*

a. Banded Argiope, *Argiope trifasciata*   b. Humped orb weaver, *Araneus gemmoides*

EGG SACS OF ORB WEAVERS

PLATE XX

B. J. Kaston

ORB WEB OF *Metepeira*

Plate XXI

George Elwood Jenks

Feather-foot spider *Uloborus diversus*, with egg sac

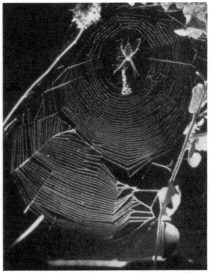

Walker Van Riper

Assymmetrical orb web of banded Argiope, *Argiope trifasciata*

George Elwood Jenks

Female of tuberculata Cyclosa, *Cyclosa turbinata*, on egg string

SPIDERS AND WEBS

Plate XXII

B. J. Kaston

ORB WEB OF *Zygiella x-notata*

Plate XXIII

Lynwood Chase

MUD DAUBER, MUD NEST AND SPIDER PREY

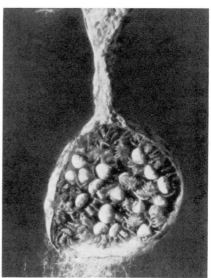

Lee Passmore   *George Elwood Jenks*

a. Portrait   b. The pendent egg sac, opened to show young

THE BOLAS SPIDER, *Mastophora cornigera*

Plate XXIV

B. J. Kaston

THE BOWL AND DOILY WEB OF *Frontinella communis*

Plate XXV

Martin H. Muma

Grass spider, Agelenopsis. An immature male sits in its tunnel

Lee Passmore

A fisher spider, *Dolomedes scriptus*

Martin H. Muma

Web of a grass spider, *Agelenopsis,* blankets the soil

SPIDERS AND WEBS

Plate XXVI

Walker Van Riper

a. Female *Lycosa* covered with young

Edwin Way Teale

b. Portrait of male *Pardosa milvina*

L. W. Brownell

c. Turret of burrow of *Lycosa carolinensis*

WOLF SPIDERS

Plate XXVII

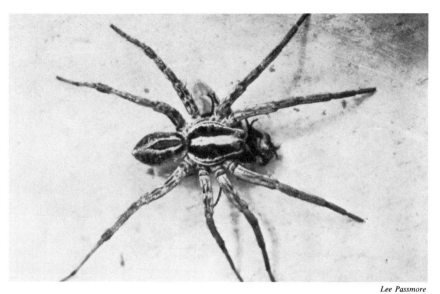

a. With captured fly

b. With attached egg sac

WOLF SPIDERS, *Schizocosa*

Plate XXVIII

a. Female with egg sac

b. Male

THE GREEN LYNX SPIDER, *Peucetia viridans*

Plate XXIX

Walker Van Riper

a. Preparing to leap

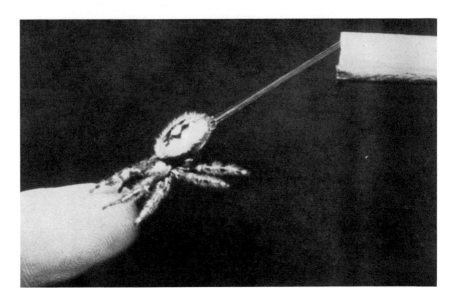

Walker Van Riper

b. Leaping

A JUMPING SPIDER, *Phidippus audax*, AND ITS DRAGLINE

Plate XXX

Walker Van Riper

a. Orienting in response to breeze, secured by dragline

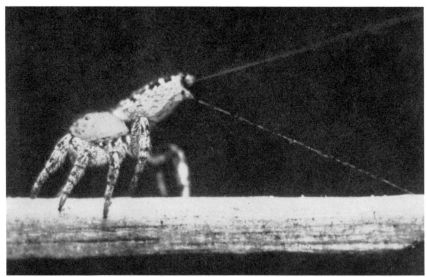

Walker Van Riper

b. Ballooning threads stream from spinnerets

A JUVENILE JUMPING SPIDER, *Phidippus*, ON A THIN TOOTHPICK,
PREPARED TO FLY

Plate XXXI

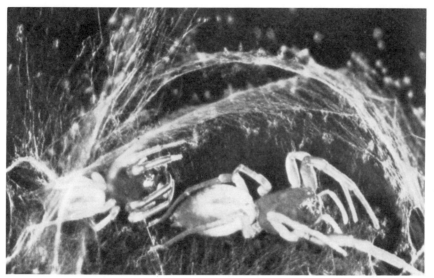

George Elwood Jenks

a. Male and female *Trachelas*, in silken cell

George Elwood Jenks

b. *Chiricanthium inclusum*, with egg sac

HUNTING SPIDERS

Plate XXXII

a. Portrait of wandering spider, *Cupiennius*

b. A wandering spider, *Ctenus*, with egg sac

c. A giant crab spider, *Olios fasciculatus*

HUNTING SPIDERS

random experimentation, during which only those suited to the minimum needs of the moment had survival value. From the first wild dragline threads laid down in haphazard fashion on and around the egg sac have evolved by progressive steps the many remarkable snares that today meet the eye. At first there were mere tangles of lines stretched without particular design, roughly filling an allotment of space between suitable supports. Probably altogether composed of dry silk, these mazes were suitable for stopping jumping or flying insects, and retarding their movements through the entangling toils. The addition of viscous drops to the lines was a later development, which transformed the stopping web into an adhesive trap. Among the lines was stationed the egg sac—the central theme, and the theoretical point from which all space webs take their origin.

The first spiders that climbed into shrubs were moving in pursuit of their favorite food, the insects that were striving to stay out of reach of their arch enemies. In leaving the soil domain these spiders were adventurers that could become full-fledged aerial types only after the web novelty had proved its worth as a means of providing food. But once the space web was in reality successful, the incessantly spinning spiders began to explore its possibilities in all directions. Some suspended a horizontal platform of rather loosely woven silk through the middle of the maze and maintained the egg sac at its core. Clinging to the underside of this, they learned to seize insects that, arrested in flight by the maze of threads, would drop to the upper surface of the sheet.

The orb web would seem to stand alone as a glorious creation, an incredible novelty designed by superior artisans. That it is only an advanced stage arrived at by the same slow steps that realized the dragline, the stopping maze, and the horizontal platform is shown in the numerous intermediate examples. The orb web is merely a formal expression of the horizontal platform. Probably at first composed wholly of dry silk, it is now provided with a large area of sticky spirals, and has been swung to a near-vertical position to make it a more effective snare. Almost invariably associated with it are some of the lines that were once the stopping maze.

The space webs exhibit a most interesting evolutionary series. Each major web type has been sponsored by different groups of aerial spiders. The primitive line weavers still rely largely on the tangle of threads for protection and as a means of stopping their prey. The comb-footed spiders spin a maze, sometimes a sheet, and almost invariably fix guy lines with sticky globules to hold their victims. The sheet-web weavers use both maze and sheets of various forms. The sticky spirals of the orb weavers hold fast an array of jumping and flying insects. Along with the webs, there have developed most interesting techniques for overpowering and enmeshing the victims; and the basic factor upon which these techniques depend is the spider's ability to move upon the web.

The successful venture on silken lines is made possible by the unpaired median claws, which lie between the much larger and longer outer pair and near their base. The median claw of aerial spiders is shaped like a hook, and is provided with a few small teeth. Associated with it are various modified hairs, which, often curved and toothed, are called "spurious" or "accessory" claws. The median claws are used almost exclusively for clinging to the lines of dry silk in their upside down position. These claws are displaced slightly to the sides, those of the first and second tarsi toward the anterior, and those of the third and fourth tarsi toward the posterior outer claw. This facilitates grasping of the threads, which fit into the hook of the claw without requiring a turning of the tarsus. When walking in the web, the spider draws the tarsi across the threads to catch the median claw, which grasps the line at an acute angle and twists it to make the grip firmer. The spurious claws orient the threads so that they can be hooked to the median claw, then act to clear the thread from the notch by uncoupling and hurling it out. With this effective device, the aerial spider moves through deep mazes or across vertical meshes with ease and precision. Its body hangs away from viscid lines that might entangle it.

Before passing on to brief sketches of the major groups of aerial spiders, some generalizations can be made that indicate the profound differences between the aerial spiders and the hunting spiders (discussed in the following chapter). Both are derived from the same primitive stocks, and on their separate roads both have become amazingly specialized. The success of each line is attested by the vast number of species found living side by side, and by the development in each series of a wide and amazing variety of types. In terms of degree of change from prototypes, the sedentary spiders have outdistanced the hunters. The silk spinners live in holes under the ground and in caves where some have become eyeless troglobites, in great numbers on or near the surface, in herbs and shrubs, and high up in trees. They have invaded space with their threads, and claim it as their own by mere placement of their three-dimensional webs. As for the vagrants, they are dominant in the soil and in the various strata of plants. They often claim space beneath the soil by digging tunnels; and the water spider has invaded the fresh water with great success. Above the ground, the vagrants move on all types of surfaces and climb into shrubbery with great agility. It may be seen, therefore, that they live side by side with the aerial spiders, but both are nevertheless to a large extent insulated from each other, almost as if they were in two worlds. King in its own domain, the hunter is usually a weakling in the clutching web of the sedentary spider. Outside its web, the sedentary spider is no fair match for the average hunter. The superiority of either line can never be tested except in terms of which one shall give rise to the dominant spider of the future. Both have accomplished great things, and stand as equals that have reached their goals by different roads.

THE PRIMITIVE LINE AND SHEET WEAVERS

The members of this group comprise a major segment of the series of spiders that took to an aerial life. They resemble in general features and equal in developmental rank the primitive hunters and weavers of the next chapter. For the most part, they are pale spiders that live in dark places, there laying down a relatively simple web of dry lines or sheets and relying on this to secure their livelihood. Most are little changed from the presumed ancestral types. The palpi and epigyna are quite simple, although in the family *Pholcidae* they appear to be specialized by the presence of numerous processes that largely mask the otherwise generalized nature of the organs. The posterior respiratory organs are tracheae. In the *Telemidae* there are two openings to the tracheal tubes, but in the other families only a single opening is present, the usual position being in front of the spinnerets. The chelicerae of the *Pholcidae* are soldered together along the midline as in the *Scytodidae* and related families, but in the other families of the series they are free. Males and females are quite similar in size and appearance; often they are found living amicably together in their webs. During the mating, the pholcids insert both palpi simultaneously into the epigynum, as do most of the primitive hunters, and the stance is the generalized type of that group. Little is known about the mating habits of other members of the series.

The line weavers of the family *Pholcidae* have globose or elongate bodies supported on exceedingly long and thin legs, a feature that causes them to be mistaken for harvestmen or daddy-longlegs. The leg tarsi are often made flexible by the presence of numerous transverse creases or sutures in the integument. Eight eyes, set close together on an elevated tubercle, are usually present, but the anterior median pair may be lost, and in some cave species all the eyes may be reduced in size or completely missing. The pholcids occupy a position between the higher sedentary types and the spiders of very primitive level. Although most common in warmer regions, they are quite numerous in temperate areas, as shown by the presence of seven genera and about forty species north of Mexico.

The pholcids spin loose, irregular webs in dark places, sometimes with a distinct closely woven sheet. Males live in the same webs as the females and resemble them closely, except for the presence of large, thickened palpi. The females carry the eggs in their chelicerae, glued together into a spherical ball and tied lightly with a few silken lines; later they may be found holding the mass of recently hatched young. Most pholcids are pale white or yellow, but some are more gaily colored in pastel greens and blues. Many are domestic and find life in buildings quite as suitable as in the open.

These long-legged line weavers are like some of the orb weavers in having a most interesting habit that becomes operative when insect prey is caught in the net. They shake the web violently to hasten thorough entanglement, then when the capture is being made, they twist the victim around and swathe it with silk. This aggressive action turns into a defensive gesture when the spider is disturbed, and it pumps up and down on its long legs so violently that it becomes a mere blur. This whirling or shuttling, which becomes increasingly violent when the stimulus is repeated three or four times, usually takes place when the web or the body of the spider is touched, but on occasion other stimuli provoke the response. When thoroughly aroused, the spider retreats to dark recesses within the web, or drops down from it to run rapidly and hide away in some dark corner.

The best known member of the family in the eastern states is the long-legged cellar spider, *Pholcus phalangioides* (Plate X), which occurs in houses almost everywhere in the world and has a wide range in the United States. A relatively large creature with a pale white, elongate body a quarter of an inch long and legs two inches long, it covers the ceiling and walls of our cellars and neglected rooms with its maze of cobweb. A very similar species with a dark pattern on the sternum, *Pholcus opilionoides*, has more recently been introduced into our northeastern states where it lives in buildings. Two of our other pholcids have been derived from outside sources: one is *Spermophora senoculata* probably from Europe but known here from the times of Hentz as *meridionalis;* and *Smeringopus elongatus*, a tropicopolitan species probably coming from the West Indies, has been taken in eastern Texas and some of the Gulf states. Only one of our species has been exported: a small pholcid from coastal California, now known as *Psilochorus simoni*, lives in buildings, especially wine cellars, from England to Italy. Many pholcids live in our western states where they live in buildings, under rocks, and in ground cavities and caves, and in suitable dark situations. Some of the largest, belonging to the genus *Physocyclus* and often living in large aggregations, have horns or outgrowths on the chelicerae of the males. On the sides of their chelicerae are fine stridulatory ridges that are activated by a coarse pick at the base of the pedipalp. Many smaller species of *Psilochorus* spin their small webs close to or seem to invade those of orb weavers, social uloborids, theridiids (Plate XII), and other spiders. They seem to be ignored by their larger neighbors so it is suggested that some sort of symbiotic relationship is involved between them. Black widows in their burrow retreats often have these spiders as companions.

In this section brief mention can be made of three families of primitive line weavers that differ from the pholcids in having six as the normal number of eyes. All are tiny creatures, rarely more than an eighth of an inch in length, and all live retiring lives in dark places under stones and debris on the ground or in caves. The sheet weavers of the family *Leptonetidae* have whitish, yellowish, or dusky bodies and hang from small tangle or sheet webs on long thin legs. Fifteen species of typical leptonetines with the posterior median eyes situated far back of the posterior lateral eyes live from West Virginia and Tennessee south into Georgia and Alabama, mainly in mountainous or deeply wooded situations of the Appalachian region. Most of these live in caves and have the eyes reduced in size; one completely eyeless troglobite, *Leptoneta georgia* (now restricted forever to its cave habitat), comes from Byers Cave in northwestern Georgia. About a dozen species of similar appearance come mostly from caves in the Edwards Plateau of middle Texas; one of these, *Leptoneta anopica* from Cobb Cave, is eyeless, and some of the others have the eyes greatly reduced in size. A third center for these spiders is on the West Coast from southern California into Oregon, and is best represented by our species *Leptoneta californica*. Four additional fairly large species with dusky bodies marked with spots and chevrons live mostly in deep ground detritus; none of these is known from caves. The most interesting of our leptonetids is *Archoleptoneta schusteri*, which is widespread in California under ground detritus, with eyes forming a transverse row and probably reminiscent of those of the ancestral stock of the entire family.

Allied to the leptonetids, but even smaller, are some cave spiders from disjunct stations in western Europe, Africa, Japan, western United States, Guatemala, and Ecuador, which completely lack book lungs and now have four orifices leading into tracheal tubes. The blind *Telema tenella* from caves in the eastern Pyrenees of France has long been the most famous representative of this family *Telemidae*. Several species, including two blind ones from caves and others from ground detritus of surface stations, occur in our northwestern states. Our best known species is *Telema gracilis*, described from Alabaster Cave, but widespread in many caves and outside situations in the middle Sierras of California. Some of our species were described in a genus *Usofila*, which is regarded as an exact synonym of *Telema*.

The third family of these primitive line weavers need be mentioned here only for completeness. The family *Ochyroceratidae* includes a number of minute six-eyed spiders from ground detritus in subtropical and tropical regions of the world. It is represented in the United States by two yet undescribed species of *Theotima* from southern Florida and the Keys.

THE COMB-FOOTED SPIDERS

The comb-footed or cobweb spiders of the family *Theridiidae* are for the most part thickset sedentary types that hang upside down from the dry threads of irregular maze webs. Most are small spiders, that suspend their snares on plants with lines so fine that they are often unnoticed, or hide them in burrows or fissures in the soil and under debris. Less well hidden are the webs of drab, house-loving *Achaearanea tepidariorum*, which soon covered with dust and debris, form the cobweb anathema of the neat housewife. Some of the most handsome and colorful members of the family are the various widow spiders of the genus

Latrodectus, whose beauty, however, is marred by an unsavory reputation. A few theridiids have hard bodies ornamented with curious spines; in others the abdomen is drawn out to amazing lengths. Most are inveterate spinners, but a few curious types (*Argyrodes* and relatives) live in the webs of other spiders as commensals, and another group (notably *Euryopis*) has forsaken a formal web for an errant life.

Most of the theridiids have rather soft, light-colored abdomens, oval or globose in form, and long, slender legs that lack spines. One of their special features is the presence, on the tarsi of the fourth pair of legs, of a line of enlarged, curved, and toothed setae that form a distinct comb used to fling silk over the prey. In most of the comb-footed spiders, the comb is strong and distinct but in the smallest ones it may be difficult to see, and in some others it has been reduced to a few weak setae. Their relatively small eyes are set close together in a group near the front of the head. Sight enters into their lives only to a limited degree, since they live in dark places and become active chiefly at night. Some males are mere pygmies beside their bulky mates, and there is often a marked sexual dimorphism. The theridiids occur in great numbers in the temperate and tropical zones; within the United States and Canada are found about 300 different species placed in nearly 30 genera. Mention can be made only of a few that typify the family or are outstanding for peculiarities of habit.

The snare of the comb-footed spider (Fig. 7, B) is not the simple mass of irregular lines that casual study would seem to indicate. It has incorporated into its limits some interesting innovations. A densely woven sheet of silk is often a feature, serving as a shelter under which the spider retreats. Leaves and debris, or grains of sand, may be used as building materials. One of the most practical homes is the bowl of the boreal *Theridion zelotypum*. Composed of dried spruce needles or other plant parts sewed together with silk, it provides a strong, waterproof tent beneath which the spider can hide its eggs and young. In some instances the theridiids leave their spherical egg sacs suspended in the scaffolding of lines in plain sight.

On the outskirts of the web at the proper season may be seen the mature males, which are received for the most part with kindness during courtship and mating. Males are killed occasionally, but not with the regularity ascribed by popular belief. The recently hatched young remain with the mother for some time, and receive consideration far beyond what one might expect from simply organized creatures. The common *Theridion sisyphium* of Europe and no doubt similar spiders from other parts of the world feed their young for several days by regurgitating fluid upon which the babies make their first meals. Thereafter for several weeks the mother and babies feed together upon insects caught and dragged into the retreat.

Many theridiid webs have, in addition to a central maze with or without a retreat, a series of longer guy lines that anchor the whole against supporting surfaces. These guy lines are held taut near their base by inconspicuous studs of viscous silk. Small insects walking against the lines are held by the glue; when their struggling breaks a line, they are lifted bodily by its contraction. The disturbance quickly brings the spider to the spot to confront the intruder.

A pictorial story of the technique used by the western widow spider *hesperus* to subdue a large wingless Jerusalem cricket (*Stenopelmatus*) is shown in Plates XXIII and XIV. The spider approaches cautiously, no doubt forewarned of the size of the prey by the strength of the pulls on the lines, then turns completely around to present its hind legs to the victim. With the aid of the comb on its flailing hind legs, it draws out heavy lines of sticky silk and ties them to the leg of the insect, until a strong band is formed. The spider next turns and injects its venom by piercing the leg with its sharp chelicerae. (Ordinarily the victim is not closely approached until completely fettered.) Then begins the task of lifting the still struggling insect off the snare floor and moving it to a suitable point in the maze. By numerous small steps, during which various threads are tightened and others put down, the bulky victim is hoisted gradually into the air until it is about three inches from the floor. Now follows the banquet. The spider feasts leisurely for three or four days on the body of the tightly bound prey; then the much shrunken remains, sucked dry, are gradually lowered beyond the inner maze and dropped to the floor.

The theridiids have long been noted for their engineering skill in lifting objects of great size. Common domestic *Achaearanea tepidariorum* is credited with having overcome and lifted small snakes, mice, and other animals. After presenting the details of captures of various small snakes by this spider, the Reverend McCook had the following to say:

> It is worthy of mention, in connection with these incidents, that the belief that a special enmity exists between spiders and serpents is very ancient. Plainy says that the spider, poised in its web, will throw itself upon the head of a serpent as it is stretched beneath the shade of a tree, and with its bite will pierce its brain. Such is the shock that the creature will hiss from time to time and then, seized with vertigo, will coil round and round, but finds itself unable to take flight or even to break the web in which it is entangled. This scene, concludes the author, only ends with the serpent's death.

Such a happening—far fetched for such small spiders—at the same time gives us some measure of the persistence of the spider and unyielding silk. One of the more spectacular feats of our *Achaearanea tepidariorum* reported by McCook was the subduing of a small mouse. About noon of a Monday it was noticed that the spider had ensnared a young mouse by spinning lines of silk around its tail. Very quickly the mouse had its forefeet on the floor and could barely touch the floor with its hind feet, and had completely failed to break the thick band of silk. By two o'clock the mouse could barely touch the floor with its forefeet; by dark the point of its nose was an inch above the floor; by nine that night the mouse was an inch and a half above the floor. By the next morning the mouse was dead.

This spectacle, watched with amazement by many people and interrupted by the clumsiness of a "meddlesome boy" who accidently broke the web (instead of by the intervention of the S.P.C.A., as is usually the case), is a compliment to the strength and elasticity of the multiple threads of the line weavers, and to their engineering prowess in elevating tremendous loads by block-and-tackle methods.

A high percentage of our comb-footed spiders belongs to the genus *Theridion* and many of these are brightly colored. The globose female of *Theridion differens*, one-eighth of an inch long with a reddish brown abdomen marked above by a red, yellow-edged stripe, places her large white egg sac in the nest. Her web is found on low plants of all kinds, and consists of a small tent, barely covering the spider, from which an irregular network of lines spreads out across the limits of the plant. More brightly colored is *Theridion frondeum* with pale white or yellow body boldly marked with black but often extremely variable in color and pattern. Some examples are almost entirely white and unmarked, whereas others have narrow dark lines or bands on the carapace and small black spots, dusky bands, or dark stripes and patches on the abdomen. These handsome theridiids, represented by one or several species almost everywhere in the United States, live on low plants and prefer moist, lightly shaded areas in woods or along streams.

Closely allied to the above theridiids are the species of *Tidarren;* our best known species, *sisyphoides*, resembles *Achaearanea tepidariorum* in size and coloration and often lives in similar situations. The males of our *Tidarren* are pygmies and bear only a single palpus, a large bulbous affair, the mate of which is extirpated just before full maturity. The female of *sisyphoides* (formerly called *fordum*) is often nearly one-third of an inch in body length, whereas the males are rarely more than one-eighteenth of an inch. At the mating season these pygmies often cluster in the webs of females, sometimes even a dozen and rarely fewer than two or three, and they seem to be tolerated by their much larger spouses.

Remarkable for their social habits are the species of *Anelosimus*, close relatives of the theridions but with more elongate bodies. Our well-known species, *studiosus*, is a light brown spider one-sixth of an inch long with dark upper and lower stripes on the abdomen. It is abundant in the South, and occurs at least as far north as New Jersey. Its small communal web is placed on shrubs and trees, and ordinarily comprises an unsightly mass of dead leaves tied together with silk and serving as a retreat, around which extends a sheet of silk attached to twigs. Several individuals live together in the nests, which except for size, are like those of other gregarious species. A very similar species, abundant in Brazil, Venezuela, and Panama is *Anelosimus eximius*, once dubbed *socialis* and well known for its social habits. Colonies of hundreds or thousands of individuals, males and females and immature stages, spin a light, transparent web, similar in texture to the sheets of the grass spiders, which has little definite form and may completely invest sizable shrubs and even trees. Some are a yard across, and are spun fourteen or fifteen feet up into the foliage of trees. The spiders wander about freely within these confines, and feed communally on insects that are captured at the periphery of the web and carried into the interior. Sometimes found in the web are such vagrant spiders as *Sergiolus* and the two-eyed *Nops;* they may be predators or perhaps symbionts, but their exact relationship to the aggregation is not known.

One group of theridiids deserves mention both for curious body forms and for commensal habits. These spiders are mostly small and, except for the vermiform types, rarely exceed one-third of an inch in length. Their legs are long and very unequal, and the tarsal claws are remarkable in that the unpaired one is long, only slightly curved, and may actually exceed the paired claws in length. The tarsal comb is reduced to three or four modified setae. Both the abdomen and cephalothorax are subject to curious variations. In one group (previously known as *Ariamnes*), the abdomen is drawn out into a long and slender cylinder, vermiform in shape, that ends in a point; in another group (formerly *Rhomphaea*), it is usually triangular in shape, sometimes extremely high, and occasionally vermiform as in the preceding genus; in *Argyrodes*, which now includes the above generic units as well, the abdomen takes many forms, being spherical, triangular, or cylindrical, and is often embellished with lumpy or pointed projections. The heads of the males of the various groups are ornamented with rounded lobes, protruding trunks, elevated spines, or other curious processes, some of which may bear the eyes.

Our single vermiform species is *fictilium*, a silvery spider occurring over much of the United States and eastern Mexico, with dusky bands on the cephalothorax and a single band down the middle of the abdomen. The body, variable in length and about one-quarter of an inch long, is sometimes triangular, but in other examples may be drawn out into a vermiform appendage one-half an inch long. This long-legged spider spins a tiny web between leaves or blades of grass, where it hangs like a straw. Its egg sac is a yellowish object shaped like a slender vase and about the same size as the spider. *Argyrodes fictilium* closely resembles some of the more typical wormlike spiders of tropical regions and like them is able to bend its abdomen back and forth. Regarding this appendage F. Pickard-Cambridge had the following to say:

This, as I have myself observed in Brazil, is wriggled to and fro, looking like a small caterpillar. But of what service to the spider this accomplishment may be is not easy to guess; for on the one hand it seems likely to attract the attention of grub-eating wasps and ants, though on the other it may attract, within striking distance, gnats and small flies who become curious to ascertain what the wriggling phenomenon may portend.

The genus *Argyrodes* comprises the multitude of commensal types known throughout the world; quite a number of species occur in the United States. All are known to spin tiny webs of their own, but they are more frequently found hanging in the webs of orb weavers, line weavers, sheet weavers, and not uncommonly in the webs of grass spiders. While hanging in these webs, legs closely drawn together against their bodies, they present an amazing resemblance to straws, twigs, scales, bits of leaves, and debris, so camouflaged that they are completely lost except to the most practiced observer. Largely immune to attack from their hosts because of small size, and perhaps also because of their cautious movement within the lines (in limited sections of which they lay down threads of their own), they feed upon tiny insects disregarded by the host. Whether these symbionts contribute to the benefit of the host is still unknown.

One of our commonest species is *Argyrodes trigonum*, a yellowish, triangular spider scarcely one-eighth of an inch long. The abdomen is high and pointed. The head of the male is ornamented with a rounded horn between the eyes and another, more slender, just in front, but that of the female remains normal. A related species is *nephilae*, a pretty black and silver spider abundant in the southern states that favors the webs of the larger orb weavers, notably those of the silk spider *Nephila*. The head of the male is produced in two long lobes, of which the upper one bears the four median eyes. A close relative called *pluto* is found in the eastern states and also lives in the webs of the black widow spider in northwestern Mexico. Another common species, *Argyrodes cancellatus*, has an elongate, triangular, gray or brownish body marked with a few silver spots and set with paired lobes on the side and at the end of the abdomen. The head of the male is produced in a rounded lobe on the clypeus, above which are two pits. This spider resembles a piece of bark or a dead leaf when lying in the web of an orb weaver or a grass spider.

Typical comb-footed weavers place complete reliance on a maze of dry lines, sticky droplets, and films from the lobed glands to ensnare insects. A few theridiids, on the other hand, have been able to divorce themselves from silk as the only means to capture prey. These spiders are small, comparatively flattened types, with legs of moderate length. They live under stones, in moss and leaf-mold, and move over the soil and vegetation with great speed. Little is known about them, but they seemingly hunt their prey as do the hunting spiders, and spin no formal webs. At least two genera from the North American fauna, *Stemmops* and *Euryopis*, belong to this series, but mention will be made only of the latter group, which is widely distributed and represented by numerous species.

The species of *Euryopis*, which resemble in a superficial way some of the crab spiders, have heart-shaped abdomens pointed behind and sometimes covered with a dorsal shield set with few or numerous long setae. Our commonest eastern species is *funebris*, a handsome blackish spider one-eighth of an inch long, whose abdomen is bordered with a silvery white stripe. Several species of similar pattern occur in the South, and prominently in the western part of our country. Another series of smaller species includes *argentea*, which is nearly black and has the abdomen pointed with four to six pairs of small silvery white spots. A similar species is *spinigerus*, orange or brown with a more distinct dorsal shield and many conspicuously curved bristles. The species of *Euryopis* are reputed to feed largely on ants but undoubtedly also prey on other small insects. On one occasion I watched a female of *Euryopis texanus* prey upon a moving line of small ants, grasping and dispatching them in numbers until the rapacious spider had gathered a small heap of victims.

THE SHEET-WEB WEAVERS

The addition of a formal horizontal platform marked a significant departure from the irregularity of the tangled space web. This strengthened zone of thin and loose webbing, with the egg sac at its hub, quickly became the theme of a new type of snare in which the upper and lower mazes and the guy lines now played a subsidiary role. The germ of the platform was present in the webs of some of the comb-footed spiders, but developed no further there; the closely woven sheet of the sheet weavers and the geometric snare of the orb weavers, however, are its direct results. These latter spiders represent a common stock that, though early branching onto separate roads, has come down to modern times as two closely allied lines. So much in common have these dominant aerial spiders that they were for a long time classified within the limits of a single large family. The comb-footed spiders diverged from the line at nearly the same time, perhaps because of failure to introduce regularity into their webs by exploiting the platform, and took a path toward perfection of lobed glands and tarsal combs.

The sheet is a yielding table upon which drop flying and jumping insects, usually after being halted in midair by a superstructure of crisscrossed lines guyed to adjacent vegetation. The sheet-web weaver clings upside down beneath the blanket, runs over the surface with rapidity, and pulls its prey through the webbing. The principal sheet acts as an effective screen against enemies from above, as well as a relatively efficient snare. A second sheet is often present beneath the hanging spider, apparently serving as a barrier to attack from below. The sheet webs (Fig. 7, A) are used for a long time; when partially destroyed by winds or falling debris, they are replaced after a few hours of spinning. In some

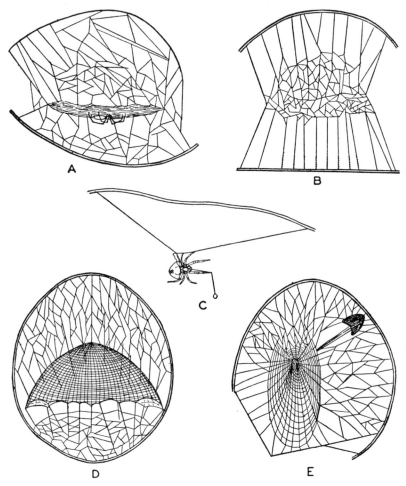

Fig. 7. Web traps of aerial spiders (schematic). A. Maze and sheet web of *Linyphia*. B. Maze and capturing lines of *Theridion*. C. Casting line and globule of *Mastophora*. D. Maze and domed orb web of *Mecynogea* (formerly *Allepeira*). E. Maze and orb web of *Metepeira*.

instances the stopping maze above the trap is missing, or is represented only by a few guy lines. Snares placed near the ground are effective in stopping and holding collembolans and small insects of many types.

The sheet-web weavers of the family *Linyphiidae* far exceed in numbers of genera and species the total for any comparable group in the temperate zones; they are the dominant aerial types. Most of the species are small, even minute, and they occur in vast, little-noticed numbers under soil debris. As a group they have more elongated bodies than the comb-footed spiders, and their legs are set with long spines. Few of them become the obese lumps so frequently found among the orb weavers; many run over the soil with a speed that belies their dependence on a fixed space web. Their chelicerae are large, strong, and well-toothed, and the straight endites are little if at all inclined around the labium. The presence of stridulating organs, most frequently a file on the side of the chelicerae and a scraping spine on the femur of the palpus, further differentiate them from the orb weavers.

Sexual dimorphism is not pronounced, except in the species with modified heads, and there is often considerable similarity in the size and coloration of the sexes. Male and female live peaceably together in the webs during the summer months.

Most linyphiids are rather plainly colored, but there are notable exceptions; some are carmine red (*Ceratinopsis*), and many have distinctive dark patterns on light bodies. As is true of most sedentary spiders, especially of those that live in dark situations, the eyes are small and little used, if at all, for the location and capture of prey. Linyphiids for the most part prefer the shade, consequently they live hidden away in dark places under natural debris on the ground, or beneath the leaves of living trees. Many species dwell in caves or animal burrows, and have in certain instance partially or completely lost their eyes.

The linyphiids are divided into two principal groups, which, quite distinct in their extremes but completely bridged by intermediate forms, are placed by many araneologists in separate families. The first of these, the *Linyphiinae*, includes the largest species, as well as numerous spinners of beautiful webs. As regards physical characteristics, the pedipalp of the female usually retains the tarsal claw, and the male palpus lacks tibial apophyses. In general the legs are longer and thinner and are set with more numerous spines than those of the *Erigoninae* (the other principal group); and the tibiae are almost always furnished with dorsal and lateral spines. Many small species belong in this latter group but space allows mention of only a few of them.

One of our largest and best known linyphiines is the filmy dome spider, *Prolinyphia marginata*, which abounds in temperate North America, and is also common in Europe. An elongate spider measuring one-sixth of an inch, the adult female has a dusky carapace with a paler marginal stripe, and a whitish abdomen heavily marked with dark bands and stripes. The conspicuous webs of *marginata* are often placed along paths or streams in shady, moist woods. The outstanding delicacy and beauty of the snare are fully revealed when the rays of the sun strike it. There is a maze of threads, extending in all directions and tied to adjacent vegetation, at the center of which is a domelike sheet three to five inches in diameter. The spider hangs below the apex of this dome. Flying insects strike the highest lines of the superstructure, drop among the closer threads, then upon the dome itself. There they are greeted by the spider, which pulls them through the webbing, trusses them up with additional silk lines while making the capture and afterward repairs the rent in the sheet. Sometimes the web is shaken to hasten the dropping of the prey. The lines of the maze and the sheet are slightly viscous, but the drops do not gather into the sticky globules used by the orb weavers and comb-footed spiders.

The common bowl and doily spider, *Frontinella communis*, found almost everywhere in temperate and tropical North America, is quite similar in appearance to the filmy dome spider. It spins two separate sheets in its snare (Plate XXIV), the principal one shaped like a shallow bowl under which the spider hangs, and the second one a horizontal sheet placed below the spider. A stopping web, largely filling the bowl and extending above it, is tied to twigs of low bushes. Snares characterized by a secondary sheet are spun by various other members of the group. Many smaller linyphiines tie a flat platform web among low plants and move about over the lower surface, but they drop to the ground and run away when disturbed.

A few of the linyphiines were reputed to spin no formal capturing web and to have become errant types, but this reputation may not be deserved. Our *Drapetisca alteranda* is commonly observed sitting flat against tree trunks, where it pursues its prey and around which it scurries actively when menaced. Its mottled gray and white body closely resembles the bark of aspens, birches, and beeches, on all of which the spider may be found; against such a background it is difficult to distinguish. It is probable that, like its European counterpart *socialis*, it lays down a fine, nearly invisible web of silk fibers and perhaps relies on it for detection of nearby prey.

Mention of so few linyphiines does not do justice to the large number of genera and species that go to make up the North American fauna. Such genera as *Bathyphantes*, *Lepthyphantes*, *Meioneta*, *Pityohyphantes*, and *Linyphantes* feature many known and still dozens of undescribed species. Our largest linyphiines, the several species of *Pimoa*, live in the cool canyons, around lakes, and in caves of the northwestern states. About twenty of our linyphiines are the same as those from Europe but most of these are old residents. What seems to be a recent and seemingly unwelcome immigrant is *Ostearius melanopygius*, a brownish spider with a mostly red abdomen tipped with black, probably originally endemic to New Zealand but now transplanted to England, Portugal, some Atlantic islands, and various parts of the United States. In England and Connecticut it has been considered a pest because of messy webs spun on greenhouse plants.

The second principal group of linyphiid spiders, the *Erigoninae*, or dwarf spiders, consists of small spinners that mostly live obscure lives under debris. The pedipalps of the females usually lack tarsal claws, while those of the males are typically armed with tibial apophyses. Most are shorter legged than the linyphiines, and live closer.to the soil, running over it quite actively when shaken from their tiny webs. They come to light chiefly when leaves, moss, and organic debris are sifted over a sheet by the careful collector. The erigonids are well known for their aeronautic habits; in autumn they make up a large part of the total group of flying spiders. Since most of them are less than one-tenth of an inch long, they can fly as adults quite as well as in the younger stages.

The small descriptive information given these dwarf spiders reflects an incomplete knowledge of their habits rather than their importance, inasmuch as they are represented by a large number of genera and species. A high percentage of the spiders of the Northern Hemisphere, as well as most of the hardy boreal types that penetrate far into the cold north and frequent the tops of our highest mountains, belongs to this group. Their tiny flat webs, fortified with a dense covering of viscid droplets, must reap a tremendous harvest of tiny insects to maintain such a population.

Some of the best known members of this series belong in the genus *Erigone*, which includes numerous dark brown or black spiders with smooth and shining carapaces armed on the sides with heavy teeth. The chelicerae and the pedipalpi are likewise often studded with sharp spines. Thse erigonids are frequently found along the edges of streams or lakes, where they place two-inch square webs among the grass roots or suspend them across stems over the water.

Many male erigonids have heads pitted and modified into grotesque shapes. A slender horn, somewhat thickened at the end and set with rows of stiff hairs, extends forward between the eyes of *Cornicularia*. In *Gnathonagrus unicorn* a single, long, slender horn projects from the middle of the clypeus. A rounded lobe carries the posterior median eyes of *Hypselistes florens*, and of many similar species, high above the remaining pairs. One of the most amazing of the erigonids is the European *Walckenaera acuminata*, whose eyes sit in two groups at the top and middle of a slender tower more than twice the height of the head itself. Often associated with these bizarre modifications are deep, conical pits usually placed just back of the posterior lateral eyes. The use to which such pits are put is known for the European *Hypomma bituberculata*. W. S. Bristowe noted that during mating the female seized the male by the head and inserted the claws of her chelicerae into the rounded pits. This observation suggests that many other species with pitted heads may perform in a similar manner, and further, that the head modifications, even though pits are absent, may be associated with interesting copulatory routines. The conclusion that modified heads and pits have arisen quite recently and independently in various groups of erigonids is supported by the close relationship with species that do not have these secondary sexual characters.

Some of the most interesting members of the *Erigoninae* have become confirmed cave spiders, troglobites, species that now cannot exist successfully outside of their particular cave or cave system. Probably our best example is *Anthrobia mammouthia*, a small white spider found in the Mammoth and other caves of Kentucky and adjacent states, which has lost all traces of eyes. It lives under stones in the caves, there spinning small flattened egg sacs that contain a few unusually large eggs. Another tiny species, *Phanetta subterranea*, is a characteristic feature of cave systems from Virginia west to Oklahoma. All its eyes are usually present, but frequently they are much reduced in size, have the anterior median pair missing, or may be completely eyeless. Since these spiders have never been taken outside of caves, it is presumed that they are troglobites quite recently restricted to cave habitats by changed climate conditions. Quite recently the late Wilton Ivie described four species of the genus *Islandiana* from caves in Virginia, West Virginia, Kentucky, and northern Texas: these pallid species have the eyes either degenerate or absent and all are classified as troglobites.

In final summary of the *Erigoninae* it must be admitted that the present treatment does not do justice to one of the most important assemblages of species in our fauna. More than forty of our species are also found over much of Holarctica, some ranging from England to Greenland. Many of these are old residents that had access across the Bering Strait land bridge and were further aided by ballooning. Others have been brought in by trade, are still being brought in, and may eventually become part of our fauna.

The cave spiders of the family *Nesticidae* resemble the theridiids in appearance, and have a somewhat similar comb of toothed bristles on the hind tarsi, but their mouth parts and genital organs ally them with the series of sheet-web spinners. Their webs are loosely meshed sheets and tangles hung on the walls of caves or hidden under stones. Some of the species are more darkly colored and live under stones and ground detritus in moist situations. The nesticid females drag globose egg sacs around with them, attached to the spinnerets in the same manner as wolf spiders. The dark recesses of caves, mines, tunnels, and cellars are the ideal habitats for these mostly pale spiders with long legs.

Three nesticid genera occur in the United States. One is represented by the single species *Gaucelmus augustinus*, first based on specimens from the cellars of the old fort in St. Augustine, Florida, and named for that city. An orange brown spider with gray abdomen, about one-third of an inch long, *augustinus* hangs from the walls of caves and only rarely is taken in dark outside situations. The species ranges from Florida through the Gulf States into Texas and eastern Mexico. Our second genus, *Eidmannella*, is represented by two similar species, mostly whitish in color and about one-seventh of an inch long. One of these is *pallida* (formerly placed in *Nesticus*), which is found over most of temperate North America, where it lives under stones and ground objects, in burrows and other dark situations, and most especially in caves. Outdoor specimens have normal, well-pigmented eyes and darker bodies, whereas some of the cave dwellers have lost some of the eyes or are completely blind. *Eidmannella pallida* has recently been found in England, where it lives under stones and in hothouses.

Our third genus is *Nesticus* and it is represented by more than twenty, mostly undescribed species. One of these is an immigrant from Europe, *Nesticus cellulanus*, which is established in a few of our northeastern states, where it lives in buildings. A principal center of our endemic *Nesticus* is in Appalachia, ranging from Indiana and Virginia south through the mountains into Alabama. *Nesticus carteri* lives in Carter Cave in Indiana and others in adjacent states but also occurs in suitable mesic situations outside of them. Most of the more than twenty species so far noted are cavernicoles and at least four of them are blind troglobites. This Appalachian fauna occurs nowhere outside of the limits noted above even though excellent cave habitats nearby seem quite suitable. Our other group of endemic *Nesticus* is found from central California north into British Columbia. One of the three known species, *Nesticus potterius*, is blind and comes from Potter Creek Cave in Shasta County, California. These northwestern species are more closely allied to those species of Japan than those of Appalachia.

The pirate spiders of the family *Mimetidae* are curious aerial types that never use silk for a snare but creep into the webs of other spiders and kill them. These handsome cannibals, once thought to feed exclusively on spiders, are now known to eat insects but mostly as a rarity. Their bodies are delicately marked with dark lines and spots. A principal feature is the presence of a series of very long, regularly spaced spines, with smaller spines between, forming a rake on the metatarsi and tarsi of the front pairs of legs. Other structural details would seem to ally the pirates either with the sheet-weaving linyphiids or the orb weavers, but their virtual failure to use silk in any way keeps their position obscure. In many respects they resemble the enigmatic *Archaeidae* from Baltic amber deposits, modern species of which have been discovered in South Africa, Australia, New Zealand, and southern South America.

About a dozen mimetids occur in the United States. Typical situations are ground debris, vegetation and, of course, the webs of other spiders. The species of *Mimetus* are about one-fourth of an inch long, and have rounded abdomens with two angled humps above the base. The species of *Ero* are about half as large, shaped like *Mimetus*, and have small humps on the top of the abdomen, with a covering of stiff brown hairs. The egg sac of *Ero* is a spherical bag covered with a loose network of brownish silk; this is twisted to form a thread by which the bag is suspended above the ground. B. J. Kaston has reported that an egg sac of *Mimetus hesperus* was made of loose threads, bright orange in color, and hung from an inch long thread.

All the mimetids are slow-moving, stealthy cannibals that have become experts in their nefarious trade. *Mimetus* preys on orb weavers and comb-footed spiders, and in the South and West is frequently found in the webs of *Tidarren sisyphoides. Ero* attacks and subdues its prey with an expertness that belies the animal's seeming innocence. This small pirate will craftily enter the tangled lines of *Theridion's* web, and clear a space of threads without making its presence known to the occupant. When all is prepared, *Ero* pulls at the lines, then awaits the approach of the aroused spinner, which hurries to the spot with customary confidence. At just the right moment, *Ero* grasps the legs and body of *Theridion* with its long front legs, and holding on firmly with the coarse rake of spines, quickly bites the femur of the victim's front leg. A complete collapse of *Theridion*, the consequence of a remarkably virulent venom, is almost instantaneous and the victor immediately begins sucking the body juices from the bulky prey. Only on rare occasions are the tables turned, and the pirate made a victim of its own seduction.

THE ORB WEAVERS

The two-dimensional snare known as the orb web is a crowning achievement of the aerial spiders. To the layman the web is an engineering triumph, a fixed geometrical object that symbolizes *spider* and partially allays unreasoning distrust of the spinning creatures. To poets of all times the orb, divorced from the spinner itself, is a celestial creation founded on beauty, its graceful spirals symbolic of the heavens and its mystery, its fragile lines a measure of the evanescence of life. To the evolutionist, it is only the last step of a series that has resulted in a circular design—an inevitable shape; and the spider has no more to do with spinning such a symmetrical web than "a crystal has with being regular." The orb web, among all objects produced by lesser creatures an unrivaled masterpiece, is above everything a superb snare. Contemplating it, one echoes the words of the meditative Fabre: "What refinement of art for a mess of Flies!"

The orb web, quickly strung up and as quickly replaced when defective, brings to the trapper an abundance of the choicest flying insects. It exploits a food supply that is active both by day and night and, in the adult winged state, available only by chance to other aerial spinners. Almost invisible in ordinary light, the lines stretch across space as a tough but yielding net into which fliers blunder, to be held by sticky, elastic threads that make the most powerful wings ineffective. (That a similar trap, produced by a like series of instinctive actions, should have evolved among a separate line of spiders might well seem an impossibility. Nevertheless, the cribellate uloborids have fashioned a web that, except for substitution of the hackled band for the beaded spiral lines, is a faithful reproduction of the snare of the orb weavers.)

This most highly evolved of all aerial webs is the result of the random activities of aerial prototypes, which finally established order among the irregular lines in the horizontal platform. During most of its history, the flat snare was enclosed in the original maze of crossed threads. At first the lines of the platform intersected haphazardly to form an irregular framework made of dry dragline silk spun from the same glands as in modern forms. Over this skeleton was laid a covering silk produced by different glands and dispensed through the posterior spinnerets, with which were mixed draglines from the ever active front spinnerets. These two elements, the framework and the covering, remain discrete throughout the evolution of the aerial flat snares.

This definite pattern underlies the sheet web of the linyphiids, even though the finished sheet may not appear to be based on a definite plan. These weavers begin at the center of the dome, put down straight lines an inch or two long, then cross them by overspinning shorter lines. These principal framework lines need only to be lengthened, and they become the radii of the orb weaver, which likewise puts down its rays from the hub outward. The primitive radii were numerous, closely spaced, and probably frequently branched so that the interval between adjacent radii at the edge of the web was little greater than that near the center. (The silk spiders still use this device to produce their strong net webs.) All these framework lines were originally dry draglines, and remain dry in all spiders. The webbing that crossed the radii was at first dry or very slightly viscous, a condition reflecting both the presence of only small amounts of sticky silk and the failure to concentrate it in heavier drops.

In the early orb weavers, the webbing over the dry framework corresponded to the viscid spiral of the higher orb weavers. The tremendous accomplishment that it represented was the formalizing of an irregular maze into a series of regular lines crossing the radii at nearly right angles. The first regular lines were probably series of curves that covered a sector of the whole, then larger loops occupying half the circle, and finally complete spirals that produced the relatively symmetrical orb web. These lines may well have been long dry rods covered with a viscous coating. At this time the formal round platform, entirely enclosed within the maze of crisscrossed threads, was still only a platform on which insects dropped. By slow stages the accompanying mazes, especially the one above the platform, were lost, but only in the highest orb weavers are they gone rather completely—and even here their vestiges may still be seen in the tangle that leads to the retreat and the hidden egg sac. The gradual inclination of the orb, and the final near vertical position, were inevitable refinements.

The evolution of the orb web progressed hand in hand with changes in the silk and in the spiders themselves. The silk glands gradually became a voluminous part of the abdominal contents, and were able to produce silks of differing properties. (In some modern forms there are more than 600 separate glands producing 5 different kinds of silk.) Viscid silk was manufactured in larger quantities and, when concentrated on the spiral lines, changed the round platform from a stopping net to an adhesive snare. As these early spinners developed, various groups branched off the main line to become sidetracked at different development levels. Some come down as probable replicas of early spiders, and their webs are significant as indicating intermediate stages. The basilica spider *Mecynogea* (formerly *Allepeira*) (Fig. 7, D), still entirely encloses its web within a maze of threads. The labyrinth spiders, *Metepeira*, largely preserve the lower maze as a tangle placed behind the orb, in which the spider rests.

The spiders that inherited the tradition of the formal orb trap comprise a multitudinous group, of which many are familiar because of large size, bizarre form, and bright coloration. They are especially noticeable during the fall months, when the orbs, and the spiders themselves, attain maximum size and cover the vegetation in great profusion. Many of the spinners are fat little animals that hang serenely in the hubs of their webs, upside down, claws pulling the rays taut, poised to move in the direction of any disturbance. Others are less bold; they sit in the comparative security of a leafy nest, but they are attentive to the thread that communicates with the center of their snare. All are accomplished trapeze artists, and swing across the lines with grace and precision. They have produced many types of orb webs; but while their success must be largely attributed to the perfection of this trap, they have also sacrificed much to gain their preeminence among the space-web spinners.

They resemble the linyphiids rather completely in fundamental features. The cephalothorax is lower and wider in front; the eyes, invariably small and little used, lie near the front edge. The chelicerae are large and strong, and the endites are short and parallel, never pointed inward. The legs may be long and well spined, but they are frequently quite short and stout. They lack the stridulating organs present on the chelicerae of the linyphiids. Sexual dimorphism is often very pronounced. In many instances, small males are quite safe within the bounds of the female's web, but not infrequently she is an ogress.

The developmental history of the orb web is only vaguely indicated in the spinning of modern orb weavers, which retain the essential details as an instinctive racial memory. The baby spider weaves its remarkably symmetrical web soon after leaving the egg sac and thereafter, throughout its lifetime, modifies the plan only in minor ways. The spinning of the orb web (Fig. 8) is an involved process consisting of a series of separate steps; to these steps most species adhere rather faithfully but there may be variations accountable to the site and weather conditions. The spider must first delimit the prospective area of operation by framing it with silken lines. The first and most important line is a more or less horizontal bridge (Fig. 8, A) on which the whole web is hung. There are two ways of establishing this bridge line. A thread may be emitted from the spinnerets and floated in the air until it catches on some object; whereupon it is pulled taut. The spider then crawls along this line, paying out a heavier thread as

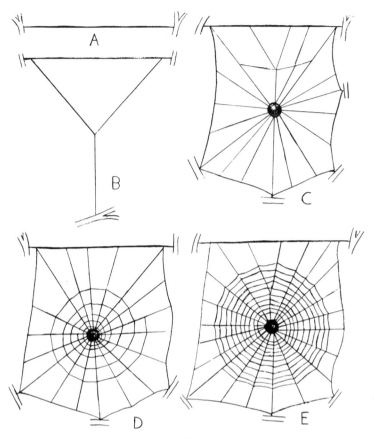

Fig. 8. Spinning the orb web (diagrammatic). A. The bridge line suspends the snare. B. The first fork delimits the hub of the snare. C. The radii are laid down in the laterally expanded framework. D. A dry scaffolding thread spirals out from the hub. E. Then viscid coils are laid down from the outside and the scaffolding thread removed.

it proceeds which will become the true bridge line, either at the expense of the floated line, which is often rolled up and eaten or discarded, or may be allowed to strengthen the new one. Another way of establishing the bridge line is by carrying it by hand down one side of the area to be covered, across and up the other side to the eventual point of attachment, holding the line free of entanglement all the time. Once the bridge line is fixed, it may be strengthened by additional threads as the spider moves back and forth across it. The bridge line is the most important thread of the orb web, often spanning wide spaces between trees or bushes and across streams, and it may be used many times to hang new webs.

The next step is construction of a frame in which to place the capturing threads. The spider first makes an attachment at one side of the bridge line, then trails out a new free line, which droops behind as she moves to the other side of the bridge and affixes it. Moving back to a near central point of the drooping thread, she then makes an attachment and drops free, trailing out a plumb line until she reaches a substratum. There she pulls it taut and attaches it to grass, twigs, or any available substratum. Thus is formed a Y-shaped figure— the first fork—at the center of which (Fig. 8, B) will be the hub of the new web while the three lines become the primary radii. The radii attached to the bridge line may be subequal in length but the lower plumb line can vary from inches to a nearby twig to a ten-foot line running from a roof corner to a flower bed below. The primary radii outline three triangular segments which, though often subequal, can vary widely in size. Into each triangular segment of the first fork, the upper one usually first, the spider now puts down additional radial and frame threads much in the way that she accomplished the first fork: by trailing out free lines, attaching them, then moving to the other side, and pulling them taut. A typical but not invariable figure at this state finds the three primary radial lines supplemented by three secondary radii in the triangles held taut by additional frame threads. The basic frame thus formed may be roughly triangular but may become polygonal by placement of other frame threads in response to local conditions and the particular habits of the spider.

Upon this basic frame of lines the spider now lays down the tertiary radii (Fig. 8, C), which may be relatively few or a great many. These radii are as a rule laid alternately on opposite sides of the enclosed space but there is often much irregularity in this practice. The tension on the lines, no doubt susceptible to testing by the moving spider, may well influence it to setting down the radii, which are sometimes placed with remarkable accuracy to form nearly equal angles at the center, but in other cases are grossly assymmetrical.

After fixing the last radius, the spider usually goes to the point where the lines converge and strengthens it by spinning a mesh of lines termed the *hub*. A part of the hub is made while the radii are being placed and stretched, and it is completed after the last one is fixed. Around the hub are then spun several spiral turns, which, because they are laid down and pulled to form irregular notches, are termed the "notched zone"; they serve to strengthen the central area and tighten the radii. The next step is to put down across the whole series of radii a spiral thread (Fig. 8, D) that holds them in place during the subsequent spinning. The turns of the "scaffolding spiral" are wide apart. Up to this point all the lines (the bridge, frame, radii, hub, and scaffolding spiral) are of dry silk. They are the framework of the primitive platform.

Beginning at the outer margin beyond the scaffolding spiral, the spider now puts down viscid spirals (Fig. 8, E) upon the dry web skeleton. It is guided, to some extent at least, by the scaffold, by tensions in the lines, by the nearness of outside lines, and by various other factors. Before the full circular turns begin, the spinning may be directed to filling in corners with short curves connecting few radii, or with longer loops that sometimes swing halfway around. As the spider gradually approaches the center, the dry lines of the scaffolding spiral are bitten out, wound up, and discarded, or sometimes eaten. The viscid spirals and loops, ordinarily composed of a single coiled thread from the beginning point to the end near the center of the orb, are placed down with a slow and deliberate motion in a very special manner. The spirals may be spun clockwise or counterclockwise.

These sticky spirals are composite; they consist of a dry core of two closely joined threads covered evenly on the outside with a viscous film derived from different glands. Following attachment of the compound line to a radius, it is grasped by the claws of one hind leg and these, as the spider's body swings across the space, prepare to fix the thread on the next radius. At the same time, the line is grasped near the middle by the claw of the other hind leg, stretched rapidly to half again its length, then let go with a snap. The result is an elastic line that contracts to the width of the space between the radii; upon magnification, it is shown to be beaded with a series of small round drops of sticky silk. The stretching of the silk line breaks up the viscid film and distributes it along the line like the beads of a necklace.

The web is now essentially finished, and the spider returns to the hub, often to alter it in the way characteristic of its group, which consists in biting out the center or ornamenting it in various ways by adding distinctive bands of silk. These bands, decorative loops, or zigzags of thick, white, flossy texture, presumably strengthen the center of the orb; from this attribute they have received the name "stabilimenta." They seem to be vestiges of the early custom of overspinning the central portion to provide a resting space in which may have been slung the egg sac. The habit is largely lost by modern types, but some of the weavers have revived and expanded it to produce striking patterns over considerable areas of the web surface. There can be little doubt that these stabilimenta, along which the spider lies or aligns its legs, give protection from various enemies.

Recapitulating, it can be seen that the finished orb web consists of various distinctive parts. A strong, moderately elastic framework, strengthened by overspinning some lines one or more times, holds in proper tension a series of radial lines of dry silk. At the center is a thickened or meshed hub, with or without extensive bands of silk, being the strategic center of the web, the place where the spider often hangs with its claws touching the radii. Beyond the hub is usually a free zone devoid of spiral lines; then comes the series of sticky spiral coils that act to snare prey. The spider hangs downward or away from the web, even when it is nearly vertical, and moves by grasping the dry lines with tiny claws. Further, the spider is anointed with an oil that to some degree prevents the sticky lines from adhering to it.

When an insect becomes entangled in the lines, the whole web is agitated by the struggles, and the vibrations are communicated to the spider. Quickly it swings across the web to the site of the disturbance, directed by the pulls on the lines, all the while trailing behind it a dragline thread, on which it can drop to the ground or save a fall if brushed from the snare. Its front legs tap the prey, further informing the spider of the nature of the victim, and bringing on a response commensurate with the problem of subduing it. Small, weak insects are seized and quickly enwrapped, but larger and more active prey are treated with greater caution. The prey is seized, held by some of the legs and turned round and round, while it is trussed with silk. Jets of fine filaments, thrown out alternately by the spider's hind legs, are combed over the insect and envelop it like a mummy. The bite may be administered either during the capture or later when the prey is carried back to the hub and feeding begins.

The struggling victim often cuts and entangles many spiral segments and radii, and may cause whole sectors to sag. During the capture, broken lines are tied together with threads by the spider, which deftly grasps the lines with its claws and pulls them together around the rent. This effective repair of the web is carried out at other times as well, by some species; still others allow the snare to become a shambles.

Mention of web repair always brings to mind the views of Henri Fabre and others who regarded spiders as creatures that, while spinning their webs, must run through the same series of inflexible and instinctive actions from start to finish. For them the spider can build another web with the pattern and peculiarities of its clan, but cannot repeat an earlier step out of its turn, cannot repair a rent in the lines. By cutting some of the lines it can be shown that the mechanical spider often spins blindly to produce an imperfect caricature of a snare. Such is also the case for spiders that are fed various drugs in testing procedures. However, spiders seem not to be quite the automatons that Fabre's view demands and that a certain amount of spinning readjustment, this also quite instinctive, is among their talents.

Many higher orb weavers spin a new web almost every night or early morning. Retaining the foundation lines intact, they remove the ragged threads and lay down fresh radii and capturing spirals. It is at this time that they frequently eat the rolled up balls of silk. Their webs are prepared for a few captures during an appropriate time interval; consequently, there is little incentive to repair a web that will not be used again. On the other hand, some orb weavers are erratic in their web construction and move from their oval retreats into the same web night after night even though it seems much tattered by overuse. The spinning of a typical orb web ordinarily requires less than an hour. Among those spiders that use the snare for a longer period are the well-known silk spiders of the genus *Nephila;* these spiders replace only a part, usually one half of their large web at one time. In other instances, there may be quite formal repair by replacement of a series of loops, or by keeping intact areas of loops and adding only the spiral or complete turns.

The principal groups of orb weavers are well represented in the temperate region of North America. Much of what we know about their natural history must be credited to the energy of the Reverend Henry C. McCook, whose fascinating and comprehensive three-volume work, *American Spiders and Their Spinning-work*, is one of the classics of arachnology. Superbly illustrated and still mostly authoritative, this is a primer to which layman and spider specialist alike can refer for dramatic essays on our orb-weaving spiders. That all ecribellate orb weavers are related as a single series is quite likely but they are now placed in several distinct families. Some modern types represent lines early detached from the mainstream and now stand as separate families.

The big-jawed spiders of the family *Tetragnathidae* are in certain features the most generalized of all orb weavers. Most are elongated spiders with very long, thin legs; the chelicerae are of great size, especially those of the males, which often project forward in a horizontal position. During mating, the chelicerae of the female are gripped in those of the male by means of long spurs that clamp the fangs and, thus firmly hooked, are rendered impotent. Most tetragnathids live in grassy areas and are especially common on the border of swamps and along streams. Some place their snares horizontally over and close to the water, but more frequently they are inclined or vertical, and framed in grass or shrubs. The snare, which has an open hub, is quite delicate, and is best suited to the capture of midges, mosquitoes, crane flies, and other small insects with weak flight.

The stilt spiders of the genus *Tetragnatha*, which appress their slender bodies and legs closely against stems or hang as inanimate straws in the center of their webs, are the best known members of this series. When disturbed, they drop on their draglines, often to the surface of the water over which they stride like aquatic bugs. A dozen or more species occur in our fauna, many of them widely distributed and abundant. One of the largest is the half-inch long *Tetragnatha elongata*, a grayish stick spider with great jaws longer than its carapace. Even commoner is *Tetragnatha laboriosa*, a smaller, yellowish species with a silvery abdomen, which lives in grass, often in dry areas.

The thick-jawed spiders of the genus *Pachygnatha* resemble the stilt spiders, but their chelicerae are shorter and heavier, inclined downward, and their legs are shorter. They live near the soil in deep grass or under debris in damp places, cattail swamps being especially favored. The young are reputed to spin a usable web but the adults wander in search of small insect prey, as do the shortsighted vagrant spiders. They live mostly in the North Temperate Zone, and are replaced southward by smaller, globose species that still use an orb web as a means of capturing insects. These latter belong in the genus *Glenognatha*, and differ in having the single tracheal spiracle advanced far in front of the spinnerets. Our best known species is *Glenognatha foxi*, an eighth-inch long, pink and silver spider of quite globose shape, which lives in meadows and grassy situations all over the South. Its delicate orb web, three or four inches across, is usually found anchored in a horizontal position about two inches above the ground, tied to grasses and weeds. A much larger species, *Glenognatha emertoni*, one-fourth of an inch long, lives in the Chiricahua Mountains and other isolated ranges in southern Arizona.

A very high percentage of the larger orb-weaving spiders of North America belong to the family *Araneidae*, a name which is preferred and replaces the long used name *Argiopidae*. The araneids are a heterogeneous lot with a number of distinctive subfamilies. Only a few colorful representatives of the five principal subfamilies can be mentioned in our limited space.

The members of the subfamily *Metinae* show some relationship to the tetragnathids, but they are more advanced in their structural features and far more diversified in their general appearance. All of them spin the orb snare, but leave the hub open; some of their webs are remarkable atypical creations. The species of *Leucauge* are brilliantly colored spiders, green and silvery white, often spotted with gold, orange, or copper; and our commonest species, *venusta*, well merits its name of beautiful. The species of *Meta*, whose best known member is the half-inch long, brown and yellow *Meta menardi* (common in Europe and long ago introduced into this country), approach the typical orb weavers in body form. Their inclined webs are placed in dark places, often in shallow caves, and the large, snow-white egg sacs are suspended by a short thread from nearby walls.

Our most interesting member of the *Metinae* is the basilica spider, *Mecynogea basilica* (formerly placed in the genus *Allepeira*), a moderately elongate creature one-third of an inch long, whose cylindrical abdomen is furnished with a hump on each side near the base. It much resembles *Leucauge*. Its web is of especial interest, since it is to a large extent intermediate between the sheet web of the linyphiids and the typical orb web. Often placed deep in well-shaded spaces in bushes, this snare consists of a large maze of intersecting lines that include a light sheet web and additional irregular lines. The dome is an orb web, constructed of a large number of closely spaced radii and crossed by a spiral line of presumably viscid silk, that has been pulled and guyed into dome shape.

Largest of all orb weavers are the silk spiders of the subfamily *Nephilinae*, exotic dwellers of the tropics, whose bodies are often more than two inches long and whose thin legs sometimes span eight inches. Their great round webs of golden silk, which will run over three feet in diameter and are supported by lines of great length, are found spanning forest paths or hanging high in trees. The giant female is attended by pygmy males scarely longer than her cephalothorax, which, although almost too small to be acceptable as food, must still cautiously approach her only after preliminary tweaking of the web threads. People who walk along paths in jungle areas frequently stride into the tough lines before they see them. Small birds are easily ensnared, and quickly make their plight hopeless by their struggles, which bind the many lines together into strong bonds. The use to which the silk of the silk spider has been put by primitive peoples and its failure as commercial silk have already been noted.

Our only silk spider is *Nephila clavipes* (Plate 18), a long-bodied, olive-brown species with lighter spots on the abdomen and long legs provided with thick brushes of conspicuous black hairs. Now largely confined to the extreme southern states, this *Nephila* was probably the same one that lived much farther north at Florissant, Colorado, during Oligocene times. The bodies of the older females are a full inch or even longer but those from the tropics are often far larger. The quarter-inch-long male makes up less than 1 percent of the female's bulk.

The orb webs of *Nephila* are not replaced as frequently as those of other orb weavers, since they have features that make them far more permanent. The dry spiral scaffolding line is looped back and forth, and is a permanent part of the web. The radii are numerous; they are branched so that the interval at the outside of the web is little more than near the center. The viscid spirals are loops for the most part, only rarely complete circles. The hub is eccentric, and is located high up near the side of the web.

In the subfamily *Argiopinae* are handsome orb weavers second only to the silk spiders in size, and far better known, especially in the North, where their bright colors and large webs make them conspicuous creatures. The argiopines differ from most of our orb weavers in having the posterior eye row strongly curved backward. The typical web of *Argiope*, the principal genus and the only one mentioned here, is ordinarily somewhat inclined but may be nearly vertical. It is provided with a sheeted hub. Frequently, the web is accompanied by a tangle of lines behind the orb, the so-called "barrier web," and less occasionally by a thinner tangle in front. These are vestiges of the stopping mazes of the primitive orb weavers; as was the case with these prototypes, they provide a protective screen against some enemies. At the center of the web is a stabilimentum consisting of a zigzag band of white silk in nearly vertical position, usually occupying a third of the diameter of the web. In some instances it is vaguely indicated, but ordinarily it is a conspicuous mark, a signature of this group of spinners. The spider hangs at the hub, head downward as usual, with the legs at each side of the stabilimentum; by appropriate stimulation it can be induced to shake the web vigorously (as do the long-legged cellar spiders and many orb weavers) until it becomes an indistinct blur. The males are very much smaller, about one-fourth as long, and mature much earlier than do their mates. In midsummer they are often found in tiny, imperfect webs near the snares of immature females; later they lurk in the threads of the barrier webs of mature, greatly enlarged females.

About six species reach the United States but only three, these widely distributed in both North and South America, are well known here. The genus *Argiope* is strongly developed in the Oriental and Australian regions with many handsome species; only a small number have been able to penetrate into the New World. The silver argiope, *Argiope argentata* (Plate XIX), is a common and characteristic spider of the American tropics, and reaches to our southern states, where it is locally abundant from Florida to California. The center of its web is provided with a two-banded stabilimentum forming a cross of white silk; the spider, mostly metallic silver and yellow, with the abdomen divided behind into rounded lobes, lies with legs stretched out in pairs to cover this zigzag cross.

Orange *Argiope aurantia* (Plate 1; Plate XVI), the females of which have bodies running to more than an inch in length, is mostly black; the abdomen, with a pair of low humps at the base, is marked above with bright yellow or orange spots. The legs of young females are conspicuously ringed in black and white, but in adults they are usually all black. The large webs, often two feet in diameter, are placed upon shrubs or herbaceous plants along roadsides, in gardens, and around houses, also in meadows and marshes. The spiders usually remain in the center of their webs even during the hottest and sunniest days. Many flying and jumping insects are captured in the snare, but a favorite food is grasshoppers, which abound on the web sites. The large, pear-shaped egg sacs (whose spinning has been described in Chapter 3) can be seen tied to shrubs in the fall or early spring.

The banded argiope, *Argiope trifasciata* (Plate 13; Plate XV), is said to be cosmopolitan and ranges widely over much of the temperate and subtropical world. It rivals the previous species in abundance but not in size. The abdomen is evenly rounded, oval in shape, without humps, and usually silvery white or yellowish and crossed by narrow, darker lines. A very beautiful spider, *trifasciata* lives in essentially the same locations as *aurantia*, but is far more abundant in grassland and drier situations. The webs are very similar, and placed to entrap the same kinds of insects. The egg sac is cup-shaped, with a flattened top.

The spiny-bodied spiders of the subfamily *Gasteracanthinae* are brightly colored creatures whose hard, leathery abdomens are ornamented with prominent spines. The spinnerets are located at the tip of a conspicuous elevation. Several of these spiders occur in the United States, but most live in the tropics, where an amazing array of bizarre types have developed. The genus *Gasteracantha* is poorly represented in the Americas; only four or five extremely variable species are found. The genus *Micrathena*, comprising more than a hundred long-bodied species, with flat or elevated abdomens bordered by long spines or thick spurs, is exclusively American.

The spiny-bodied spiders hang on short legs in the centers of their webs, looking rather like chips of wood, bits of leaf, or plant fruits. The sharp spines make them unpleasant morsels for birds, lizards, and vertebrate animals, but their worst enemies, the solitary wasps, fill mud cells with them, not at all deterred by the armor. *Gasteracantha's* webs are inclined or vertical, and have open hubs. The radii or foundation lines are ornamented with a series of flocculent tufts of whitish silk. It has been suggested that these may serve as lures for midge-eating insects which, deceived by the white flecks, might fly into the orb and be caught.

Two species of *Gasteracantha* are found in the United States, but one of these, the mainly West Indian *tetracantha*, is very rare. The common *Gasteracantha cancriformis* (Plate 16), of our southern states is subject to considerable variation. It has a yellowish or orange abdomen spotted with black and fringed by six black spines.

Our species of *Micrathena* are equally spectacular in their bright coloration and curious shapes. *Micrathena sagittata*, an arrow-shaped species having a white or yellow abdomen armed with tiny basal and median pairs, and at the apex a greatly enlarged pair, of divergent red-tipped spines, is common even in our northeastern states. Lumpy *Micrathena gracilis*, a yellowish to brown spider, whose elevated abdomen is set with five pairs of short spines, is representative of a quite different series. These, and several others not mentioned here, have small males that in shape do not closely resemble their females.

The typical orb weavers of the subfamily *Araneinae* far outnumber those of other subfamilies and in the temperate regions they are the dominant group. Mention can be made of only a few of the more than twenty genera, some with scores of species, that occur in our fauna. The physical characteristics of most araneines are a generally thick-set appearance, bulky abdomens and relatively short legs, but some have become elongate types that can run quite rapidly. The abdomens of these typical orb weavers are subject to very considerable variation in shape. Some are leathery, and surmounted by humps or spines that make them resemble some of the spiny-bodied spiders, from which they differ in not having the spinnerets on the end of a tubular eminence. Sexual dimorphism is pronounced in many genera, perhaps reaching its acme in the bolas spiders; in certain cases the males may be essentially equal in size to the females, except for the bulk of the abdomen. It is among these orb weavers that the male seems to run the greatest risk while courting, but he has learned to reduce the danger by dropping on his dragline when his attentions are unwelcome. His long front legs are usually armed with rows of spines that aid in holding the female and in keeping her at arm's length when she pursues him. With the exception of the few species that have modified their ensnaring habits, the typical orb weavers (including *Eriophora ravilla;* Plate 14) spin a round web, but considerable differences are found in the details of the orb and accompanying retreats.

The small, elongated species of *Cyclosa* (Plate XXI), usually about one-fourth of an inch long, have conical humps at the end of the abdomen. Through the center of their beautiful snares, furnished with many radii and closely set spirals, often lies a stabilimentum consisting largely of the remains of insects and debris tied together with silk. The eggs are later added to the string. The spider sits at the hub, bridging the space between the segments of the string and blending so completely with it as to be practically invisible.

The webs of most orb weavers still maintain, at least in vestigial form, the ancient mazes of their prototypes. In the labyrinth spider, *Metepeira* (Fig. 5, E; Plate XX), the maze has been retained as a prominent, irregular net, which the spider uses as a base and in which it hangs its leaf retreat and string of eggs. Many species occur in the United States of which *Metepeira labyrinthea* widespread in the eastern part is representative; in size they vary from one-fifth to one-half inch in length. Their orb web is usually complete, with several trap lines leading from the hub to the retreat of the spider in the tangle. The spiderlings use the labyrinth as a nursery after they break out of the egg sacs, and it is reported that they sometimes feed upon small insects caught in the tangle.

Several of the typical araneines habitually spin incomplete orb webs, entirely omitting the spiral lines and radii from a segment equal to the space between two or three radii. This they accomplish by spinning rounded loops, and swinging back and forth many times instead of making complete circles. Associated with such snares is a trap line, sometimes virtually bisecting the open sector but usually in a different plane, that leads to a nearby retreat. Our several introduced species of *Zygiella* (Plate XXII) spin incomplete orbs of this type, but better known to Americans are those of *Araneus pegnia* and *thaddeus*, which have a wider distribution. The lattice spider, *Araneus thaddeus*, is about one-fourth inch long and has a rounded abdomen of pale, yellowish color with darker sides. Its beautiful silken retreat is usually attached to a nearby leaf; inside the retreat lies the spider, holding the trap line stretched to the center of the orb. Spiders that use these trap lines, when activated by an insect catch, first swing to the center of the web, then directly toward the point where the prey is entangled.

A great many of our largest orb-web weavers spin a complete orb and communicate with it by means of a trap line stretched from a retreat of folded leaves. Typical is the common shamrock spider, *Araneus trifolium* (Plate 15). This pretty spider, often more than one-half inch long, has an evenly rounded, white to pink abdomen, usually marked above with a three-lobed spot resembling a shamrock. Its carapace is banded with brown, and its legs are conspicuously ringed with white and brown. The male is only about one-fifth of an inch long. Even more strikingly colored is the related round-shouldered *Araneus marmoreus*, common over most of the United States and Canada and well known in Europe, which has a bright orange body and an abdomen marked by contrasting darker lines of an indistinct folium.

Larger even than the round-shouldered orb weavers are some that have their bulky abdomens produced into prominent basal humps. These gray or brown spiders sit at the side of their complete orb webs in a crevice, under chips of wood or bark, or in a more formal leafy retreat. Many of these araneines measure one-half to a full inch long and seem to favor coniferous forests particularly in our northern states and Canada. In the western states one of the most familiar is *Araneus gemmoides* (Plate XIX), a pale species that ranges eastward into the Great Lakes states. Females vary in size from one-fourth to three-fourths of an inch, but the males are only half as large. The male of the even larger relative, *Araneus illaudatus*, which lives in southeastern Arizona and west Texas, is only one-sixth of an inch long. A common darker species of this series, *Araneus nordmanni*, has a dark folium on the abdomen. Although most common in North America, this species also occurs sporadically in northern Europe. One of the most interesting species of this group, *Araneus diadematus*, was introduced into this country by trade and now lives in our northeastern area from New England into the Great Lakes states and in a more recently established area on the West Coast from Oregon to British Columbia. This species, called the "cross spider" in Europe, favors domestic situations around houses and city gardens. *Araneus diadematus*, easy to maintain in the laboratory, has long been a favored experimental animal in Europe; in the United States Dr. Peter Witt has established that different drugs fed to this spider produce distinctive patterns in their web building.

Included among the typical araneines is one small group that has repudiated the orb web in favor of a distinctive and extraordinary method of capturing insects. These are the bolas spiders of the genus *Mastophora* (Plate 17; Plate XVIII), fat creatures mostly one-half inch long, whose bodies are ornamented in a most grotesque manner. The carapace is bedecked with sharp, branched crests or horns, and set with many small, rounded projections; the voluminous abdomen is lined and wrinkled and surmounted with rounded humps. These bizarre specializations, reminiscent of similar ornamentation in the dinosaurs and other groups of animals, are not known to play an important part in the life of the spiders.

The hunting site of female *Mastophora* is usually the outer branch of a shrub or tree, most often well off the ground. On this branch the spider hangs in plain sight. Mementos of her previous activities are numerous silken lines that soon form a thin coating over the twigs and leaves. Hanging to the lines, or hidden among the nearby leaves, may be one or more egg sacs, beautifully and durably made and featuring a long, coarse stem drawn off from the globular base. During the daylight hours *Mastophora* clings to a twig or leaf, completely immobile, perhaps deriving some protection from her resemblance to various inanimate objects. Even when handled, she shows only momentary evidence of life, and may be rolled around in cupped hands like a marble. Few spiders are so completely inscrutable.

But *Mastophora* is a creature of the evening and night, and as we watch her then in the performance of her marvelous routine, we forgive her earlier listlessness. The disappearance of the last rays of twilight is the signal for action; she takes up her position for the evening's sport.

With plump body swinging from the ends of her legs, she moves to the end of a branch and affixes her thread to the lower side by pressing her spinnerets against the bark. Grasping this thread with one of her hind legs and holding it away from the branch, she crawls several inches farther along and pastes the other end firmly in place. The result is a loosely hung, horizontal line which she often strengthens with an additional dragline thread. This strong trapeze line is hung far enough below the branch to allow a clear space for casting. Moving now to the center of this line, *Mastophora* touches her spinnerets and pulls out, to a length of about two inches, a new thread that lies clear of the other. Keeping it attached to her spinnerets and held taut, she combs out upon it quantities of viscid silk. Each hind leg alternates in producing the liquid, until a shining globule as large as a small bead is formed. The spider now pulls this line out still farther, allowing the weighted portion to drop part of the distance to its natural point of equilibrium, then she turns and severs it just below the globule with the claws of one of her hind legs. The freed line swings back and forth like a pendulum, but the spider turns quickly and approaches it, searching and groping with her front legs until she is able to grasp it. Quickly she swings her massive body and seizes the trapeze line by the hind legs of one side, adjusting the casting line between her palpi and one of her long front legs (Plate 17). Poised and ready now is the bolas spider, waiting—with the patience that characterizes her—for the approach of a suitable victim.

Also aroused to activity at this time are many nocturnal insects, which soon fly along accustomed lanes, dipping down close to the foliage and fluttering in and out among the branches. A large-bodied moth, its wings spreading nearly two inches, its great eyes shining red in the last rays of reflected light, dips down toward the hunting grounds of the spider. As the moth approaches, *Mastophora* gives every evidence of knowing that a prospective victim is near. She moves her body and adjusts her line, as if in tense expectancy. At just the right moment, when the moth comes within the reach of the line, the spider swings rapidly forward in the direction of the flier. The viscid ball strikes on the underside of a fore wing, and brings the moth to an abrupt stop, tethered by an unyielding line which will stretch one-fifth of its length before breaking.

Fluttering furiously at the sticky end of the lasso, the moth makes every effort to free itself, but the spider is quickly on hand to give the final coup. She bites her victim on some part of its body. With the venomous bite resistance ends quickly; and the paralyzed moth can be rotated and trussed up like a mummy in sheets of silk. *Mastophora* then sets to work feeding on the body juices of her catch. This bountiful food supply will keep the spider busy for some time. After having satisfied her appetite, she cuts the shrunken remnant loose from the trapeze line and drops it to the ground below. Later in the night a second capture may be made, but *Mastophora's* needs for food are usually well met with a single sizable victim.

It must not be concluded that the life of this spider is quite as simple as the incident protrayed might indicate. *Mastophora* may wait in vain for a flying creature to come near enough for capture. In many instances, her aim may not be as accurate as pictured or the prospective victim may be too large to be held even by the strong band of silk. But patience is one thing at which spiders excel, and *Mastophora* is no exception. Should no victim reward her after half an hour of waiting, she winds up the globule and line into a ball and eats it. Quickly she spins another line, prepares another sticky bead, and resumes her vigil. Some students believe that aerial pheromones attract the moth to the spider but for that thesis there is at present no evidence.

The capturing device of *Mastophora* represents the heritage of a small tribe of spiders found in various parts of the world. Heber A. Longman of Brisbane, Australia, described the similar angling methods of *Dicrostichus* and dubbed it "angler spider." In South Africa, Conrad Akerman described the innovative technique of *Cladomelea:* the line with sticky globule is held by the short third leg and whirled rapidly with a rotary motion in a horizontal plane, as is a whirligig. After about fifteen minutes of strenuous whirling, *Cladomelea* pauses to rest and renews the line for a new effort at capturing a victim. Close relatives of our bolas spiders are found in many other parts of the world and it is presumed that their capturing techniques will be found to be similar.

How wonderfully complex is the pattern of instinctive actions that makes up the casting habit of *Mastophora!* Although endowed with glands that produce silk in copious quantities, she bases her whole economy on a blob of sticky silk dangling at the end of a short line. And still not content with a niggardly use of this vital material, she eats the viscid globule if it is not put to use. The trapeze line, the pendulum thread, the viscid globule, and the instincts of a hungry spider have combined to produce one of the most sensational of all devices for the capture of prey.

Can this parsimonious use of silk by spiders dedicated to spinning of the orb web be carried even farther? A phlegmatic neighbor of *Dicrostichus* is the orchard spider *Celaenia*, which rests upon a branch fully exposed to view without any attempt of concealment. *Celaenia* spins silk on her tree retreat, lays egg masses near her, but fails to prepare a capturing web of any kind. A fluttering moth approaches the retreat, seems to be guided to the inanimate spider, and when near enough is grasped by the spiny legs of the spider. It has been suggested that some pheromone allures the moth to approach and become victim of the raptorial spider. On the other hand, it is possible that the flight of the moth is a random one and the success of the spider based on unlimited patience.

We now leave the typical araneines for a quick review of various small, globose orb weavers that have diverged rather sharply from the more prosaic members of the group and seem to lie in the vague intermediate zone between the three principal families of aerial spiders. The ray spiders of the family *Theridiosomatidae* have as a typical representative the widely distributed *Theridiosoma radiosum* of our eastern states. Females of this ray spider run about one-tenth of an inch long and have rounded, oval, highly arched abdomens marked with many small silvery spots. The remarkable egg sac, a brownish, pear-shaped bag, is suspended by a long, often forked thread from the branches of trees or the sides of stones. The aerial station makes it immune to depredation by crawling insects.

The ray spider is commonest in dark, damp situations and favors shaded woods, the underbrush along streams, and nooks at the base of cliffs. The web, usually vertical in position and three to five inches in diameter, is a most remarkable structure. It is first spun as a reasonably typical orb web with meshed hub and several spiral turns in the notched zone; then these threads are bitten out. The radii, a dozen or so, are next rearranged so that they converge upon a small number of lines that radiate from a point at or near the center. These rays, in turn, converge upon a short trap line that is attached to a convenient twig or surface. The spider rests its body on the rays of the orb and, sitting upright and facing away from the snare while holding the slack line loosely between its front legs, pulls the web into the shape of a cone or funnel—or an umbrella turned inside out, as McCook described it. When an insect strikes the web, the spider lets go the line and both snare and spider spring back to aid in the further entanglement of the victim. Only one of the rays, comprising three or four radii, is badly damaged with each capture; so the spider uses the trap several times.

The upright position is most unusual for an aerial spider. It is made possible by the spider sitting upon a foot basket of taut lines and clinging with its hind legs. The resemblance of this device to that of the triangle spider is most striking. *Hyptiotes* uses a single sector, hangs back-downward from the ray threads, and springs forward on the line when the trap is released. A single capture destroys the triangular snare, but *Theridiosoma* has several sectors in reserve.

The small members of the family *Anapidae* share some unusual characters with the following family. They are lungless since the front pair of book lungs has been transformed into tracheae. The pedipalps of the female have been lost. At least some of them are known to and others are presumed to spin minute orb webs. The only anapid so far described from the United States, *Chasmocephalon shantzi*, is a tiny, subglobose, eight-eyed species, about one-twentieth of an inch long, with a leathery abdomen covered with numerous tiny brown rings. It is widespread in California and Oregon, where it lives in ground detritus and other dark situations, including caves in the Sierras.

The minute spiders of the family *Symphytognathidae* are relatives of the anapids, also lack book lungs and the females have aborted the pedipalps, and they have six or four eyes in diads. Our only representative is *Anapistula secreta*, a tiny, globose, four-eyed spider of widespread range in North America south of our borders, which has recently been taken from ground debris in southern Florida.

A third group of these small to minute spiders share some features of the above two families but they are now given full family status as the *Mysmenidae*. Some of these have typical book lungs but others are intermediate or completely lungless. The pedipalps of the females are present and of normal size. Some of these spiders are reputed to spin small, irregular webs but others are known to spin minute orb webs. It is likely that all were derived from orb-weaving forebears.

Several mysmenid genera occur in the United States and the group is strongly developed south of our borders. *Trogloneta paradoxum*, a blackish species in which the carapace of the male is prominently elevated, lives in ground debris from Utah to California and Oregon. *Mysmena incredula*, one of our smallest spiders only about one-thirtieth of an inch long, has the suboval abdomen marked with about eight small whitish spots. This pygmy lives in ground debris from Texas to Florida and in the Bahama Islands. The many species of *Maymena* in Mexico spin tiny orb webs on the walls of caves; our only species, *Maymena ambita*, lives in ground debris in our southeastern states and has often been taken in caves. *Lucarachne palpalis* of Honduras has been found living in the webs of the diplurid *Ischnothele* seemingly as a commensal. The spiders moved about in all parts of the web and numerous females carrying egg sacs were noticed.

10
The Hunting Spiders

The hunting spiders are for the most part bold creatures that put only moderate reliance on silk to gain a livelihood and spend much of their life in the field. They run upright on the soil and on vegetation, and maintain this upright attitude even when on webs. Among them are conspicuous extroverts, whose open ways have earned them such names as "wolf" and "fisher" spiders, "running" and "jumping" spiders, and other names that describe quite suitably the characteristics of animals that pursue and overpower their prey by strength, speed, and alertness. Their strong, usually elongate and cylindrical bodies are propelled by stout legs of moderate length, as befits running creatures. Many are big-eyed hunters with keen sight that stalk their prey during the daytime. But at the same time we also find numerous allies of retiring, even secretive habits—shortsighted vagabonds that skulk under the dark security of ground debris, that come out only at night to grapple fiercely with small creatures touched by their groping legs.

The line of the vagrants starts with the same shy, shortsighted ancestral spider that gave rise to the aerial sedentary types. Originally far less venturesome than its modern cousins, this prototype retired to the cover of a stone or a crevice, where it deposited its eggs and enclosed them in silken sheets. Then around itself and the precious bag it spun a silken tube or cell, at first left open at both ends, later closed behind, or in front as well. In this comparative security it spent most of its time, departing only for short hunting forays, after which it dragged the prey back to be devoured at leisure. Allegiance to silk was a moderate one. Dragline threads were put down during the foraging; and the elementary subservience to these lines still remains firmly fixed in the habits of the boldest and swiftest of the vagabonds. Silk was used by the males for sperm webs, and invariably by the females to make the flattened egg sacs, which were composed of lower and upper sheets joined at the margins.

Few of the hunters have completely given up the silken cell as a base. Some found such comfort there that they have remained in it throughout their history, and have modified it only by embellishing the entrance with various types of webs. *Ariadna* lies in the tube and waits for insects to trip on the signal lines that radiate from the mouth. Many funnel-web spiders spin a little silken collar around the opening and await callers with like patience. The well-hidden funnel of the grass spiders provides a sanctuary from the gate of which the spider may survey its vast sheet web where drop jumping or flying insects. All these spiders are fundamentally hunting types; they represent a very distinct line from those creatures that use the third claw as a hook to swing through space. The sedentary vagrants rarely produce aerial webs of consequence, and they emulate only poorly the superb devices of the aerial snarers.

Most of the early hunting spiders found it advantageous to move away from the bonds of silk. Improvement in vision made possible a life of action far from the retreat even during the daytime, and some were molded into swift vagrants, with little need for a fixed station. Two distinct lines have been followed by the higher hunting spiders; one culminates in the wolf and lynx spiders, and the other in the jumping spiders.

Whereas the curved, unpaired claws were the prime determinants for the departure of the aerial spiders from the main line of spider evolution, these had little to do in laying down the path of the vagrants. The wolf spiders and their kin retain the median claw, but it is small in size and not used as a hook. No claw tufts or accessory claws are ever present, but in some of the heavy ground forms the lower surface and sides of the distal leg segments are covered with thick pads of hairs. The gradual development of better eyesight in their prototypes made possible longer and longer forays away from the cell, thus leaving the egg sac vulnerable to the attacks of predators. During the egg-laying season the females remained near their sacs to guard them from depredation, and some remained on hand until the progeny had emerged and dispersed. The lynx spiders and funnel-web spiders still guard their eggs in this fashion. Other spiders learned to mold the flattened egg sac into a round ball and carry it around with them held by the mouth parts beneath the body. The fisher spiders still use this cumbersome method. Both the stationary vigil and the unwieldly ball put strong restraint on the normally active lives of these spiders. To ease the curb, some transferred the rounded sac to their spinnerets so that it could be dragged, a position that permitted normal hunting. This habit is the badge of the wolf spiders.

In the remaining vagrant line, the unpaired claw has been lost and the tarsi are supplied with adhesive claw tufts that allow the spiders to climb with great ease. In some of the wandering ctenids—*Cupiennius* and its relatives—the fading median claw can still be seen beneath the claw tufts; it serves to bridge the gap between the three-clawed and two-clawed vagrants. The type reaches its acme in the big-eyed jumping spiders, which are the most alert and in many ways the most highly developed of all spiders. Another principal branch has been the series of laterigrade families culminating in the typical crab spiders. Finally, at the very base of the series are the six-eyed hunting spiders, the remnant of an ancient group that has retained many primitive features.

THE WOLF SPIDERS

The handsome wolf spiders of the family *Lycosidae* are expert hunters that have few peers among the vagrants, and among all araneids are excelled only by the jumping spiders. They occupy almost all terrestrial habitats and seem to be at home in all as dominant predators. Some are amphibious types that rarely stray far from water, skating over or diving under the surface when they are menaced. Others have become adapted for a secretive life in areas of shifting, open sands, into which they dig tunnels and on the surface of which they hunt during the night hours. Most numerous in prairie regions, the wolf spiders abound wherever a plentiful insect food supply is available among the grasses, and where the sunshine penetrates all but the densest clumps.

Many wolf spiders have deserted their heriditary silk-lined cell for a life in the sun. Others, more conservative, return periodically to the retreat; some pass much of their lives there, leaving it only to hunt. Quite a few have improved the retreat by changing it into a deep tunnel in the soil, this in certain instances being closed by a movable trap door. Only one group of wolf spiders, *Sosippus* and its relatives, has moved in the other direction—that is to say, toward a greater dependence on silk; it spins a sheet web similar to that of the grass spider.

Except for mere size, which varies widely between tiny quarter-inch *Piratas* and giant *Lycosas*, an inch and a half or more long, there is a surprising similarity in appearance among the wolf spiders. The elongate cephalothorax is usually high and narrowed in front, and bears eight eyes whose size and position immediately distinguish the lycosids from almost all other spiders. Set close together on the lower part of the face is a row of four small eyes that point forward and slightly to each side. Immediately above these eyes are two very large eyes that point forward, and farther back on the dorsal part of the head are two large eyes that look upward. The spider is thus able to see in four directions, and because of the size and acute vision of some of these batteries, can perceive moving animals at a distance of several inches. The legs and chelicerae are robust, as befits such powerful hunting creatures; the oval abdomen is of moderate size.

The capture of prey by the wolf spider is marked by vigor and power. The spider pounces upon its victim and, holding the body in its strong front legs, bites and crushes it with its stout chelicerae. The capabilities of this rapacious hunter are not without limitation, however, when contrasted with those of higher animals. Although keen and longsighted among spiders, its vision is far inferior to that of most insects. Its prey is perceived by sight but the character of the moving object is probably not at all evident until the spider touches it. The diurnal lycosids are undoubtedly able to make greater use of their eyes than the nocturnal types, but these latter are conditioned to respond to the slightest disturbance of the soil of their hunting ground. Furthermore, the wolf spiders have a tapetum that reflects light rays back through the eye retina, and this presumably improves their night vision.

The female wolf spider is a creature of variable temper. Notorious for her rapacious activities, she nevertheless displays a solicitude for her eggs and young (Plates XXVI, XXVII) that can scarcely be matched by any other spider. The mother *Pardosa*, which it will be recalled encloses her eggs in a carefully molded subspherical bag, attaches the sac to her spinnerets and drags it around with her wherever she goes. It makes no difference that it is often as large as she is; this egg bag is a precious thing to her and she will defend it with her very life and fight viciously to retain it. Her instincts are most powerful ones, but—ironically— she is easily fooled and will accept for a time, and almost without question, a substitute sac from which the eggs have been pilfered, a piece of cork, or a wad of paper or cotton of the proper size and shape.

After two or three weeks, her young develop to a point where they can leave their crowded quarters. The mother then bites open the sac at the seam, and within a few hours a whole brood of tiny spiderlings has climbed upon her back and huddled there in a mass. The cluster will completely cover her abdomen and much of her carapace, and very often is composed of more than one layer of spiderlings. It seems to be true that the *mother* must open the egg bag, and that without her assistance the babies often perish.

During the time of carrying the young, the mother engages in normal hunting activities, and her children must accommodate themselves to a strenuous life. She will run with great speed when pursuing or being pursued, turn to defend herself when cornered, and during all these wild gyrations the spiderlings cling to her back. When brushed off, they quickly crawl back upon their perch, if they have the opportunity. During this waiting period they do not take food, a fact that has led to considerable speculation as to how they are able to survive. By some they were thought to derive energy from the sun and air. However, adult spiders are notorious for their ability to go without food, and the spiderlings are equally tolerant. Their bodies are provided with a food supply, and this is adequate to maintain them until they start feeding. While they are riding on their mother's back, which may be for a full week, they are biding their time until the next molt, after which they will leave her to take up separate lives in the grass, and begin their own hunting activities. The spiderlings do drink water during the ride, and probably find a sufficient supply in the dewy film that often covers them at night. They have been observed to move to water and take their fill when the mother stops to drink, then clamber back on her abdomen.

The success of the wolf spiders in surviving is unquestionably due in part to the initial protection given to the eggs and the young by the mother; she contributes by maintaining a vigil over the sac, by carrying it always with her, and by seeing that it is opened at the proper time so the young can emerge. Thereafter, however, the clustered spiderlings seem to remain with her through their inclination rather than hers. She pays little attention to them, and abandons them if they fall off and cannot again reach her of their own initiative.

It is possible to give in this brief section only a glimpse into the lives of a few American wolf spiders. A wealthy fauna awaits the enthusiast who cares to investigate them further.

The wolf spiders of the genus *Pardosa* are small, but they make up in abundance what they lack in size. In physical appearance they feature large eyes occupying nearly the entire width of the head, which is quite precipitous on the side. More gracefully built than the typical lycosid, they have a slender body supported by long, thin legs set with long spines. Their slender tarsi lack for the most part the conspicuous brushes of the larger lycosids. Their colors tend to be dark, frequently black, but the cephalothorax is usually marked above by a pale longitudinal stripe continuous with a light band on the abdomen. The heads and forelegs of the somewhat smaller males are often brightly variegated with white and black patches of hairs, features which are displayed during courtship activities.

The species of *Pardosa* are true vagrants and do not use a formal retreat for long, wandering instead over the soil and low vegetation in moist areas. All are sun-loving creatures and abound in the spring, at which time the males become mature and cavort in front of the more plainly colored females. Except in the far north, where more time may be necessary to complete full development, they live only one year; in the case of males, months less. Noted for their excessive agility, they climb into flowers and over plants, and the spiderlings are often seen ballooning in the fall.

Dozens of species of *Pardosa* live in temperate North America and occupy many different habitats. The moss- and lichen-covered slopes of the far North and the highest mountains support distinctive dark species. In the dried grasses of meadows and along roadsides live small species striped in black and gray. In the Southwest, rocks and bare sands in and along creeks serve as stations for speckled species that are hardly visible when not in motion. Most *Pardosa* abound in damp, grassy situations near bodies of water and some are amphibious, being able to run over the water freely and to crawl under the surface by holding on to plant stems. One of the British species, *Pardosa purbeckensis*, lives in the intertidal zone and takes to the water during high tides, in the manner described by W. S. Bristowe:

> The following day was sunny and a lot of the spiders were actively running about, but as the tide rose, they retreated to the higher portions of the plants.

Presently I saw one which I had been watching touch the water several times, like a bather feeling the temperature with his toe before taking the plunge, and then it deliberately walked down the stem of the plant beneath the surface, taking with it a bubble of air, caught by means of its hairy body. I watched several others, and the same thing occurred, and this is therefore, how they survive the high tide. I was puzzled at first by seeing that they dived long before they were forced to by the submergence of their plant, but this was explained by an individual that got dislodged, for it could not dive without the help of something firm to hold on to, and even the tip of a leaf swaying in the current was not sufficient aid. Although they can run over the surface, they are far more comfortable beneath it, especially in the rough weather, so the wisdom of their submerging whilst something firm remains to cling to becomes clear.

Pardosa is a small lycosid. There are some smaller—"pirates" of the swamp-loving genus *Pirata*, and the shy *Trabaea* of shaded woods—but in the main the typical wolf spider is larger and more stoutly built, and will often attain notable dimensions. Most of the typical wolves belong to the genus *Lycosa* (from the Greek meaning "wolf," or "tear like a wolf"; it is also the common name for the whole group), and strength is the keynote of their makeup. The carapace is low and the sides of the head broadly rounded, so that the eyes ordinarily do not occupy the whole top of the head, but sit in a group at the center of the dome. The rather short, heavy legs are often supplied with dense brushes of hairs beneath the tarsi and metatarsi.

These typical wolf spiders are very handsome creatures. Their bodies are evenly covered with a dense coat of black, brown, or gray hair, which gives them a velvety appearance. Paler markings of various kinds, arranged in spots, patches, and stripes, add variety to the rich coloration of the hairs. Whereas the upper part of the body tends to harmonize with the terrain, the underside of both body and legs is often boldly marked with black patches and stripes.

Lycosa's egg sac is almost always white in color. The female molds it into a nearly spherical object, and turning and spinning over the edges, leaves scarcely any evidence of the seam where the two sheets have been joined. It is in many ways a much more finished piece of work than the flattened bag of *Pardosa*, and would appear to be better adapted for dragging over the soil.

Many of these lycosids are very active day hunters. Various handsome and distinctively striped varieties abound in grassland and in grassy areas along roadsides over most of the United States. One of the best known is long-legged *Lycosa rabida*, whose gray cephalothorax is marked by two chocolate-brown stripes and whose abdomen displays a median brown stripe margined in yellow. A near relative is *punctulata* (Plate 21), in which the dark dorsal stripes are conspicuous, and the venter of the abdomen varied with a series of small black points and markings. The body of *Lycosa hentzi*, mainly a Florida species, is yellowish and resembles dried grass. All these striped wolves are good climbers and often ascend high into grass bunches and low shrubs.

One of the most interesting habits of the *Lycosidae* is the extended and highly developed tunneling practiced by certain species. The splendid burrows made by them were not, of course, perfected in a single step. We can trace their gradual evolution in the habits of their creators. At the outset the wolf spider took temporary refuge beneath a stone, and lined the area with its characteristic silken cell. But space requirements for the growing spider often made it necessary either to enlarge the cell or to abandon it. In order to use the first of these alternatives, the spider had to develop the use of its chelicerae as digging instruments, and of its silk to bind the soil so it could be removed from the premises. The primitive burrows that resulted from attempts to enlarge the living quarters were only shallow depressions in the earth immediately below the cell retreat—and many contemporary wolf spiders still dig this type of pit. But other species increased their proficiency, moved their burrows to favorable sites in the open, and dug tunnels. Some made a further improvement by erecting at the burrow's mouth an elevated turret to serve as a lookout. Developing along another line, a few lycosids have learned to cover the entrance with a movable lid similar to that of the trap-door spiders.

All the burrowing wolf spiders of the United States are relatively large spiders that live more than one year, and in some instances do not attain full sexual maturity until the second year. The spiderlings establish their burrows soon after leaving the mother, and gradually enlarge them as they grow. They dig with their chelicerae, which are not, however, provided with a rake as are those of the drap-door spiders. They tie the soil together with silk into little pellets, which they carry in their chelicerae and drop a short distance from the burrow entrance. The walls of the vertical tunnel are lined with silk, a very important material in the construction of the domicile; and the spider's movements are facilitated by a ladder of webbing that allows it to climb quickly and surely to the surface. The considerable reliance of these wolf spiders on silk is further noted in the various refinements associated with the burrow opening: the turret, the winter and summer aestivating closures, and the trap door—all are dependent on it.

The typical burrow conforms throughout most of its length to the size of its occupant, but an enlargement, usually in the middle portion, allows the spider to turn around and serves—in the exact sense of the word—as a living room. Because of cramped quarters, mating ordinarily takes place on the surface, after the males have enticed the females to come outside. As for their maternal habits, the burrowing wolves transport their egg sacs and young around with them, even while moving in and out of the narrow tunnel. They have learned to carry the sacs to the entrance, where they can be exposed to the rays of the sun; a mother will sit just inside the opening, and turn the bag over and over with her legs and palpi to warm all its surfaces. (This habit seems to be a necessity for nocturnal species and for those that scarcely move outside the entrance during their day hunting.) Whereas the vagrant wolves are usually rid of their young a week or so after they have clustered on the mother's back, the young of the burrowers may remain with their parent for long periods, sometimes over the winter in the tunnels.

Collectors seeking specimens will find that the burrowing lycosids may occasionally be duped by using a decoy—an insect tied to a string, a wad of beeswax, or a stem to which the enraged spider will cling long enough to be pulled out of its burrow. When the spider sits near or has been coaxed to the entrance by some method, a quick jab with a knife blade or heavy forceps will close the lower part of the tube and make capture easier. Digging the burrower out may prove a laborious undertaking if the tunnel is tortuous or established in rocky soil. Those lycosids that live in sand are easily taken with shovel or trowel, but it is a wise precaution to put a stem into the burrow, or fill it with dry sand, and then follow the course down. The spider will usually retreat to the narrow bottom and lie there quietly, well hidden with soil and not easily discovered until completely unearthed.

Along the margins of North American streams and in sandy fields live a number of pale species that are here termed "bank wolves." The best known of these is *Arctosa littoralis*, a grayish spider one-half inch in length that is flecked with many dusky markings, and often blends with the sand or gravel on which it sits. It is quite at home in loose sand, and frequently digs a burrow in this material, binding the grains together with silk and encircling the entrance with a collar of small stones. Our *littoralis* is widely distributed from Canada to southern Mexico. Many individuals seem not to dig any sort of burrow, and instead will be found hiding under stones along lake shores and water courses. This species also hunts along sandy beaches on our East Coast. A similar species, *Arctosa serii*, is found on the sandy beaches of the Gulf of California where it burrows in sand along the drift line and is occasionally submerged by the tides.

The largest of our wolf spiders is *Lycosa carolinensis* (Plate 21, 25), a mouse-gray spider that combines a vagrant life in the open with the more prosaic one of the burrow. Females of all ages can be found wandering about or hidden under debris, the adults often dragging their huge egg sacs or carrying their numerous young. In the North these inch-long creatures assume a uniform dark grayish brown, and the whole venter of the body is jet black. Examples from Texas and northern Mexico are far larger in size, lighter in color, and have the venter speckled or banded with black. Some specimens from Arizona and Sonora have the abdomens, especially the venters, pinkish in color. Most specimens of *carolinensis* probably burrow at some time in their lives and these burrows are most commonly encountered in open country on relatively dry hillsides and in prairies covered with a sparse growth of low plants. The upper part of the tunnel is always inclined, and the deeper part is often quite tortuous, lying among roots and stones. The entrance is large and may lack an external modification, though on occasion this great spider builds a high turret (Plate XXVI) of grasses, sticks, or stones around the hole.

A particularly interesting variant on the turret theme is that of *Lycosa aspersa*, the "tiger wolf." This handsome spider, dark brown in color and possessing stout legs marked by many pale yellowish stripes, lives in open woodland in our eastern states and digs its tunnel straight down six or seven inches into the rich humus. Around the mouth it erects a high parapet of moss and debris, and over the top of this parapet spins a canopy, leaving an opening on one side only. On top of the canopy are placed bits of soil, moss, and leaves, so that the whole nest is well hidden and blends with its site. In many instances the canopy is more than just a rigid covering; it becomes a hinged lid that may be lifted and dropped to close the opening, and in this form is comparable to the wafer doors of the true trap-door spiders.

Mary Treat was the first to describe the burrow of the tiger wolf; she observed over an extended period the life and general activities of a colony of twenty-eight of these spiders. In spite of their well-camouflaged nests, half of which were sealed during most of August, all but five tiger wolves fell victim to the digger wasps during that month. Those that escaped had completely cemented down the lids of their nests until the wasp season was over. Such a great toll seems to suggest that *aspersa* is no safer living underground than her several close relatives, which rarely dig into the soil, and then make only a shallow cavity.

In the southeastern part of the United States live many large *Lycosa* to which the common name "sand wolves" may be applied with considerable accuracy. A representative species, the most widely distributed of the whole series, is *Lycosa lenta*, a pale wolf covered evenly with grayish hairs and only lightly marked above by a dusky pattern. Intensive daytime collecting in Florida, where these sand wolves are most numerous, rarely produces examples of the several species; at night, however, under the rays of a headlamp barren areas and seemingly unproductive habitats become bejeweled with their eyes, and it is possible to capture quarts of specimens within a short time.

These wolves are abundant on white sands, where they lie quietly with their legs outspread. They have fine eyesight, but rely almost entirely on touch to capture insects. When the sand is tapped with a pair of forceps, the spider rushes over to grasp and wrestle with the instrument almost as it would with normal prey. Most intriguing of all the sand wolf's reactions occurs when it is disturbed: it turns a somersault, dives into the sand, and disappears, leaving on the smooth surface no sign of where it has gone. Careful investigation shows that there is a well-hidden burrow closed by a perfectly concealed trap door. This door is coated above by a fine layer of sand; it is very thin, even thinner than the most tenuous wafer door of the trap-door spiders but essentially similar to it. The sand wolf opens the lid quickly and crawls headfirst into the cavity, closing the door after her with her legs.

The burrowing life has left such small imprint on the bodies of its practitioners that they appear to differ in no important respects from the vagrant wolf spider. They produce subterranean burrows comparable in excellence to those of many trap-door spiders, but without benefit of the specific modifications that the latter enjoy. Only in the "earth wolves" of the genus *Geolycosa* do we find features that suggest a first step toward true adaptation to a subterranean life. Many wolf burrowers have thick, round bodies and modified appendages, but in *Geolycosa* the cephalothorax is higher and more strongly arched than usual, and the chelicerae are unusually robust. The front legs are very stout and proportionately thicker in both sexes, and all the legs lack prominent dorsal spines. The earth wolves are confirmed exponents of a subsurface existence, and spend almost all their lives within the burrow. Extremely shy, they are reluctant to move very far from the opening even when capturing insects, and usually sit partially inside, ready to retreat at the slightest distrubance. Other wolf spiders will wander a few feet from the opening to wait for prey, or even forage long and far from the tunnel retreat. These may be approached at night with a headlamp and easily captured, but the nervous earth wolves usually must be dug out of the soil.

The species of *Geolycosa* (see Plate 21) are to be found over much of the United States and temperate Mexico. Some are yellow-brown spiders clothed with whitish hairs, but most have dark reddish and brown bodies, masked by a covering of slate-gray or brown hairs. The undersides of the body and legs are usually marked with jet-black bands and spots. They dig their burrows from six to twelve inches into the ground—the depth being somewhat dependent on the character of the soil—and line the whole with silk. Ordinarily the tunnel goes almost straight down, and is enlarged in the middle or at the bottom. Some of the palest American earth wolves (such as *wrighti* and *pikei*) live in the open sand of beaches and inland dunes, while the darkest species, *rafaelana*, digs in the red soil of our southwestern deserts. Those earth wolves that live on bare surfaces ring their burrow openings with an inconspicuous collar of coarse sand grains glued together with silk. Still other species (*missouriensis* and *turricola*) are found in the plains or on hillsides where there are numerous small objects suitable for use in turret building. These spiders almost invariably erect a prominent lookout from whatever materials are close at hand, fitting the pieces together with meticulous care by bending pliable straws and pine needles to the shape required. The turret, which has been likened to an old-fashioned log cabin chimney, is bound together with silk and has a smooth inner lining continuous with the silk of the burrow. Some are similar to the nests of birds and exhibit workmanship requiring quite as much skill.

One small group of wolf spiders has given up vagrancy in favor of a sedentary existence on the top of a sheet like that spun by the grass spiders. It is generally believed that these sedentary wolves once placed only moderate reliance on silk, and that snare-spinning was acquired later in their history. The typical lycosids were probably running spiders by the time they took to hauling their egg sacs about, and the fact that the sedentary wolves still use this practice in their webs suggests that the sheet is a secondary development. These sedentary wolves differ only slightly from the typical wolf spiders. The cephalothorax is flatter; the eyes are somewhat more widely separated; the lower margin of the chelicera is usually armed with four stout teeth, instead of the three or two of most other lycosids. The legs are rather long, and the tarsi and metatarsi are so thickly covered with hairs as to form quite wide brushes, particularly on the front pairs. The posterior spinnerets are considerably longer than the anterior ones; their apical segment is prominent, and they are probably used to a considerable extent in putting down the fine and closely spun webbing of the snare.

The sedentary wolves appear to be most abundant in tropical regions but some species live in the southern part of the temperate zone. In penisular Florida *Sosippus floridanus* spins its funnel retreat under beach debris and lays its sheet over dry sand. This spider is quite dark in coloration, the deep red to brown carapace being marked with a median pale line and broader marginal stripes of white or yellowish hairs. The abdomen is dark gray above, with darker flecks on the sides, and with a broad median stripe, also low in tone, running its full length. The females average about two-thirds of an inch in length. Common *Sosippus californicus* of southern Arizona, California, and adjacent Mexico is quite similar to the Florida species but much paler, being light brown or even yellowish. It bears a close resemblance to some of the large grass spiders, and when moving over the sheets may easily be mistaken for those swift runners.

THE FISHER SPIDERS

There is one group of spiders that is rarely observed far from the moist edges of streams and lakes, and that includes some members wonderfully adapted for life near or on the water surface. These amphibians are the handsome vagabonds of the family *Pisauridae*, animals of large or even giant size that resemble the wolf spiders closely in appearance but differ from them quite distinctly in certain habits. The pisaurids broke away from the true wolves early in their history, became committed to existence in moist areas, and now seem largely limited in their distribution by the presence of permanent streams or ponds. Often referred to as "water spiders," they are no more than distant cousins of the European *Argyroneta*, a spider that has adopted the aquatic medium to such an extent that it lays exclusive claim to that title. They are also called "nursery web weavers" because of the spinning industry of the females, but the most appropriate appellation is that of "fisher spider"—a name that properly conveys their amphibious aptitude as well as their occasional fondness for little fishes.

The typical fisher spider of the genus *Dolomedes* (Plate 20; Plate XXV) is a huge gray or brown spider with an oval abdomen and a longitudinal cephalothorax more flattened than in most wolf spiders. The integument bears many appressed plumose hairs, in addition to various simple hairs and spines. The eyes have much the same arrangement as in the *Lycosidae*, but the dorsal row of four is not so strongly curved, and rarely is markedly larger than the front row. This would seem to indicate that the range and acuity of the fisher's eyes are less than those of the typical wolves. They are big-eyed hunters, nevertheless, and seem to have excellent day vision.

No obvious physical features in the bodies of the *Pisauridae* identify them as spiders of the water, but they walk over the surface with a grace nearly equal to that of the water-striding insects. The tarsal hairs are probably arranged to give buoyancy and to push them when skating, but only in some of the species do conspicuous brushes adorn their legs. Much of their success as pond skaters must be attributed to their extreme lightness, which, repudiating their physical bulk, keeps them from breaking through the surface film. Their slight weight, however, while of great advantage on the surface, becomes a liability when the situation calls for submarine action. The aquatic pisaurids cannot swim as does *Argyroneta*, and seem able to break the water surface only with great effort. Diving is impossible unless they are able to exert considerable force with their legs on some convenient support, and they remain submerged only by clinging to underwater leaves and stems—bobbing up to the surface like corks when they release their hold. The aquatic pisaurids are able to remain beneath the water for long periods. One instance of forty-five minutes has been noted and the limit is probably much longer. The body hairs capture bubbles of air, thus making the spiders even lighter; and although this further impairs their swimming abilities, some of the bubbles come in contact with the respiratory orifices and furnish the needed oxygen for their underwater sallies.

In an excellent revision of our *Dolomedes*, James Carico found eight species living within the borders of the United States and some of them vie with the giant wolf spiders for the honor of being our largest true spiders. Perhaps the largest is *Dolomedes okefenokensis*, a robust fisher spider covered with dark brown hairs mottled by lines and spots of grayish and yellowish hairs, first discovered in the Okefenokee Swamp in Georgia. The females attain a body length of more than an inch and a half, and their lightly ringed legs span four or five inches. The male is considerably smaller and somewhat more brightly colored, the pattern of dark spots and pale lines usually being quite distinct. Similar species occur commonly in the northern states and, even though of smaller size, seem formidable enough when happened upon along beaches or in boathouses. The sluggish streams of the United States, particularly those of our southeastern states, harbor many species of these vigorous creatures. *Dolomedes albineus* (Plate 20), an ash-gray species of average size, is a confirmed aquatic type and deserves special mention. It rests with legs outspread and head-downward on the trunks of cypress and tupelo trees in the southern swamplands, completely motionless unless disturbed, when it whisks out of sight around the tree trunk like a squirrel, or dashes into the water to skate away or hide under the surface.

Brief mention should be made at this point of the species of *Trechalea*, a group of large American fishers similar to *Dolomedes*, of which a single, yet undescribed, representative extends its northern range into the high mountains of southern Arizona. *Trechalea* flattens its grayish, black-flecked body against a stone at the edge or in the water of streams, poised to skate out at the first sign of a struggling insect. The activities of this creature mark it as one of our finest fishers, an expert with unusually good eyesight, and fringed tarsi that are very long and flexible to aid in water walking.

Most handsome of our American fisher spiders is *Dolomedes triton* (Plate 20), a green-gray animal of moderate size. The upper part of the cephalothorax is marked by two silvery white lines passing down each side and continuing the whole length of the abdomen. A narrow white band runs from between the eyes to far back on the cephalothorax, and the abdomen is further marked with four or five pairs of small white spots. On the sternum are six dark spots, a distinctive badge of *triton* which had given it the now rejected name *sexpunctatus*. This spider is the most truly aquatic of all our *Dolomedes* and haunts the wettest portions of swamps and streams over much of the United States. It is often seen on the water surface, its hind legs moored to the edge of a water plant and its other legs far outstretched and lightly pressed into the water film. When disturbed, it will run over the water and hide in the aquatic vegetation, and when closely pursued it clambers into the water and hides underneath leaves or debris.

Dolomedes triton is a close relative of the English *Dolomedes fimbriatus*, the "raft spider" of the Cambridgeshire fens. This handsome fisher, also common in Europe, resides in swamps and other standing or slow-water habitats and lives a life-style much like our *triton*. Her common name was based on a now rejected report that *fimbriatus* constructed "out of a few dead leaves and some threads of silk, a small raft on which it set sail on the face of the waters. From this raft it sallies forth over the water in pursuit of its prey, for it can run easily on the liquid surface." A much more appropriate name for this largest of British spiders would be "swamp spider" as suggested by W. S. Bristowe.

The food of the amphibious fishers consists mainly of the larger terrestrial insects from bank vegetation, and of aquatic insects in various stages that are found crawling in the shallow water or living on the muddy edges. On occasion, however, these predators have been seen capturing small fishes and tadpoles and feeding upon their bodies. This activity can be considered peculiar and surprising only if the preconceived notion exists that spiders must feed solely on insects and are unable to assimilate the bodies of vertebrates. The truth of the matter is that spiders rarely hesitate to attack any creature that comes within certain size limits. A tiny, squirming fish, twice the size of the spider itself, is no more formidable an opponent than a robust grasshopper, and is as easily dispatched. The spider bites with its strong mouth parts, and its venom proves very active on cold-blooded animals. Furthermore, its powerful digestive juices appear fully as effective on the bodies of fishes as on those of the invertebrates that are its habitual food.

There are a number of well-authenticated instances of the angling prowess of our North American *Dolomedes*, enough to suggest that the capture of tiny fish is not a rare occurrence. While fishing in a swampy region of the upper St. Johns river in Florida, Dr. Thomas Barbour watched the capture of small cyprinodont fish by spiders that swarmed on the floating lettuce and other vegetation.

A tiny flash of silver caught my eye, and I looked again, to see a spider carrying a small dead fish, perhaps an inch long, across a wide leaf to the dark interior of a large lettuce cluster. I thought that probably the spider had found a dead fish by chance and I relit my pipe, when about six feet away in another direction the episode was repeated. This time the little fish was still struggling feebly in the spider's chelicerae. Later, I saw a third fish being carried off which was dead and quite dry.

Some *Dolomedes* have been seen to capture trout fry in hatcheries, and are considered capable of accounting for a good number of such tiny fish. The owners of balanced aquaria have sometimes been puzzled by the disappearance of prize fish, subsequently to discover a spider robber with part of its spoils.

It is improbable that vertebrate prey forms more than a small portion of the total food of the aquatic fisher spiders. One wonders whether the toll even closely approaches the great number of spiders of all sizes eaten by trout and other surface-foraging fish. Once the fisher spider has captured its prey, often after a struggle in the water itself, it must retreat to a dry station to enjoy the victim. The predigestion and eating must be accomplished on land, inasmuch as the digestive juices of the spider would be diluted and lost in water.

The pisaurids are not all noted for their amphibious activities; many, while abundant in moist areas, do not enter the water itself. The best American example is *Pisaurina mira*, a fisher spider which ranges widely east of the Rockies and lives in open woodland often far from water. About one-half an inch long when full grown, *mira* is a very pretty spider extremely variable in color and pattern. Its most common shade is a light yellow-brown, marked the whole length of cephalothorax and abdomen by a wide darker band bordered by white. Our *mira* is frequently found on bushes and low vegetation in woods and in grasses and mixed vegetation of meadows and old fields.

Other species of the genus *Pisaurina* have become elongate spiders markedly resembling the slender crab spiders of the genus *Tibellus*. Our two species live in the southern coastal plain and lower piedmont of our southeastern states, being most abundant in Florida and the Gulf states. The first of these, *Pisaurina undulata*, has a pale yellow body, and may be marked with narrow longitudinal lines of darker color. When at rest they often lie with their long legs closely appressed to the surface of stems. This species is often associated with aquatic habitats and lives in vegetation around bodies of water. A second, similar species, *Pisaurina dubia*, has a similar range but lives in somewhat drier situations but still often near water. There are many of these slender pisaurids in Africa, and at least one genus, *Euprosthenops*, builds a funnel and sheet web similar to those of the grass spiders. The snare is often more than a square yard in extent, spreading over the branches of shrubby acacias with the funnel near the ground. *Euprosthenops* is reputed to hunt on the underside of the sheet as do the true sheet-spinning sedentary types such as *Linyphia*. By their use of silk far exceeding that of the true wolf spiders, the pisaurids are seen to intergrade with the *Agelenidae* in some of their habits as well as in structure.

The pisaurids are often cited, and with considerable justification, as providing the outstanding example of maternal devotion among all spiders. Many spiders guard their eggs for varying lengths of time, but few have made of it a complicated ritual that gives protection to the young until they are ready to scatter. Only the wolf spiders exercise equal care, and they are close allies of the pisaurids.

The first act of the pisaurid mother in behalf of the coming generation is the spinning of a silken cover around her eggs, which is at first white but usually becomes gray or even brownish. From the time it is made until the spiderlings are ready to emerge, the mother carries this treasure around with her wherever she may go, holding it between her long legs and underneath her body. The claw tips of her chelicerae are inserted in the ball, her pedipalps press around the sides in front, and silken lines from her spinnerets moor it securely from behind. It is often so large that the mother is forced to run on the tips of her tarsi in order to hold it clear of the ground. The difficulty of transporting such a tremendous object seems to be very great, and it is fortunate that the habit is operative at a time when the normal desire for food is considerably inhibited. A few of the fisher spiders transfer the egg sac to their spinnerets and drag it about as do the wolf spiders, but most of them have retained the ancient and awkward method.

Dolomedes carries the ball until just before the young are ready to emerge, or until a short time after emergence, then fastens it to a suitable spot at the end of a branch of some herbaceous shrub. A three-lobed leaf is often chosen as the site, and the leaflets are pulled down and tied with silk to form a cozy retreat, the nursery web. It may be supposed that the female aids the young to escape by opening the egg sac; thereafter the babies quickly spin their tiny lines and scatter within the confines of the nursery. The mother remains outside the retreat guarding the spiderlings until they have molted and moved away, a period often of more than a week.

Pisaurina mira (Plate 19) usually prepares her nursery well in advance even of the egg-laying. She displays a decided preference for poison ivy, using its leaflets for the top and sides of her retreat and spinning up the opening below with a platform of silk. After the eggs are laid and enclosed in a sac, *mira* hangs on the outside of the nursery until just before they hatch, at which time she suspends the bag in the nursery.

THE LYNX SPIDERS

The lynx spiders of the family *Oxyopidae* are hunters that have become specialized for a life on plants. They run over vegetation with great agility, leaping from stem to stem with a precision excelled only by the true jumping spiders. A few are more indolent, and sit in flowers or press their bodies close against dried stems while they await the appearance of suitable prey. The lynxes hunt mostly during the daytime, aided by a relatively keen eyesight comparable to that of the wolf and fisher spiders. Although they trail a dragline even when jumping, silk does not enter much into their lives, and they never make use of webs to capture their prey.

The typical lynx is a strongly built creature with a high, oval cephalothorax and a rounded abdomen tapering to a point behind. Its thin legs are all about the same length, quite long, and armed with long black spines. The tarsi always lack brushes of hairs, but the absence of such pads does not seem to detract from its climbing ability. It has dark eyes placed either in two rows so strongly curved that they seem to form a circle, or in four rows of two each. They are unequal in size, the anterior median pair being very small and some of the others quite large, as befits the spider's active, diurnal life.

The lynx spiders are best represented in warmer regions; however, about twenty species occur within the limits of the United States, and a few are common far into the north. One of the most conspicuous is the green lynx, *Peucetia viridans* (Plate 22; Plate XXVIII), which is abundant in the southern states from coast to coast, and also occurs in Mexico and Central America, where it is the commonest and most widely distributed member of its genus. The female is a large spider often three-fourths of an inch in length, and her slender mate is not far inferior in size. *Peucetia* is usually colored a bright transparent green variegated with rows of small red spots. A red patch usually adorns the face between the eyes. Rows of long, black spines are a conspicuous feature of the thin legs, which are ringed with red at the joints.

Most of the examples of *viridans* from the southeastern states are tinted the same bright green, and there may be a relationship between color and habitat, but so far this lynx has not been identified as a confirmed resident of any particular species of plant. In California, a favorite site for the spider is the dull green foliage of the wild buckwheat (*Eriogonum fasciculatum*), and the egg sacs are frequently found tied to the yellowish flowers of this woody shrub. Many of these western lynxes are yellow or even brown in color, and have the whole dorsum blotched with large red markings that often form a complete band. Some of the Old World *Peucetia* are reported to live almost exclusively on a single plant. One variety is said to frequent the fresh green tufts and to be bright green in color; whereas others that habitually seek the dried areas of the plants are yellow and strongly marked with a pattern of pinkish spots.

The straw-colored egg sac of the green lynx will be found securely lashed to the outer twigs of her plant home, and over it the patient mother hangs, head-downward, hugging the bag with her long legs. The sac is nearly as large as the spider and far more bulky—a rounded object whose thick outer covering is embellished by many small, pointed projections. From the egg sac extends a maze of lines to nearby leaves, investing the whole branch in a silken web, where the young can remain until they are ready to fend for themselves. The nest of the green lynx is often similar to that of the fisher spiders, and her maternal solicitude for the tufted purse and the young that break out of it is not less strong than in the makers of the nursery web.

The remaining lynx spiders of America are much smaller than our green *Peucetia*, but they are more numerous in species and more diversified in color and pattern. Common and representative is the striped lynx, *Oxyopes salticus*, which is at home in both the temperate and tropical zones of North and South America. This pretty spider is about one-third of an inch long. The female has a pale yellow cephalothorax clothed with white scales and varied by four longitudinal bands of dark scales. Her abdomen is mostly white and is marked above by a dark basal dash and below by a dark median band. Her pale legs have a narrow black line beneath the femora. The male is smaller, has black palpi and a black abdomen, which often possesses an iridescent sheen. Another species from this group is the gray lynx, *Oxyopes scalaris*, a brownish spider clothed uniformly with gray hairs. This common lynx has penetrated farther north than our other species and is found all over the United States, being especially abundant in the western states, where it lives on sagebrush and similar plants. The habits of these lesser lynxes are quite similar. All are plant spiders, run on low bushes and herbs, and there place their disciodal egg sacs suspended in a little web.

THE WATER SPIDER

It was pointed out earlier that the name "water spider" is reserved for *Argyroneta*, one of the most amazing of all spiders, a land creature that has taken to life in an alien medium. *Argyroneta* is not truly aquatic since she must still have air to sustain life, but she has transferred her aerial environment to a station beneath the surface of the water, and there remains for long periods. Although found only in Europe and temperate Asia, *Argyroneta* is included here because no general book on spiders would be complete without some mention of her extraordinary behavior. Another index of her prestige is placement by many in a family of her own, the *Argyronetidae*.

In appearance *Argyroneta* is a very ordinary spider about one-half an inch long, plainly clothed in dark brown raiment and unmarked by a contrasting color pattern. Nothing in her physical aspect indicates proficiency in swimming or diving; no appendages are present that might serve as effective instruments to propel her or to maintain her beneath the surface. Severely plain when outside the water, once *Argyroneta* dives she becomes a shiny, silvery bubble, transformed from a drab gnome into "an elfin fresh from fairyland."

Many spiders shun the water. Others, it has been seen, live near it all their lives, and often move over the surface or crawl beneath it to stay for short periods. *Argyroneta* is the only spider that can live entirely in the water and that is able to swim and move about without having contact with submerged objects. Most spiders are able to survive immersion for limited periods because they take a bubble of air with them, held closely to their bodies over the air spiracles. *Argyroneta* supplies her primary need for oxygen by mounting to the surface and raising her abdomen to capture an air bubble. Just how long can she stay underwater without renewing this supply? It has been calculated that if it were possible for her to lie motionless in the water, theoretically the armor of air would last about sixteen hours. However, *Argyroneta* is an active swimmer and expends her oxygen supply more quickly, making it necessary to come to the surface at frequent intervals.

The favorite haunts of the water spiders are ponds and sluggish streams in which aquatic plants are plentiful and in whose quiet waters she can best display her swimming talents. The first of her underwater domiciles is built in the spring and serves her well during the warmer portion of the year. In a suitable bower of vegetation not far below the surface *Argyroneta* lays down a platform of silk, suspending it by numerous attachments to adjacent plants. The closely woven sheet and staying lines are so like the water in color that they are quite invisible at first. This framework finished, *Argyroneta* swims to the surface for air to provision her unique home. She raises her abdomen and hind legs well above the water to secure a large air supply, then submerges; the brushes of long, curved hairs on her rigidly extended hind legs form a screen to aid in keeping the air bubble fast beneath her body. She then paddles underneath the sheet and releases the air which pushes upward and billows the silk into a small air sac. After many trips to the surface, the silk has been blown into a miniature diving bell, open below, which from the outside appears as a silvery drop in the water. There follows additional spinning on the bell and further tying with supporting stays to make the finished retreat a durable structure. To it the spider brings fresh air as the need arises. On occasion, the bell will be cut open at the top to allow air to escape, after which the rent is repaired and the air renewed.

Much of the life of the water spider is spent within the confines of this underwater chamber, where feeding, molting, mating, and rearing of the family all take place. Hunting goes on for the most part at night and the prey, consisting chiefly of small aquatic animals, is dragged into the bubble to be digested. Predigestion of the food is possible only in an aerial space provided by the bubble of air. At the time of mating, the male spins his smaller diving bell close to that of his mate, then joins the two with a silken tunnel. At other times he will omit this preliminary and swim directly to share the bell of the female. Considering the cramped quarters, it is probably just as well for the satisfactory conclusion of his suit that the male *Argyroneta* is usually larger than the object of his affections.

Long after the mating, the eggs are laid and cradled in a tough sac hung in the upper part of the bell, where they hatch after about three weeks. The spiderlings move into the spacious lower portion and remain there until they depart the nest. Expert swimmers from the beginning and equally skilled in underwater architecture, they soon fill the water with their own tiny bubbles of quicksilver.

In the fall *Argyroneta* moves into the deeper recesses of her water environment and spins another domicile which will serve her as winter quarters. Much more durably constructed than the diving bell, this home is usually a closed sac spun in the cavity of an empty snail shell or a similar shelter. During the cold months the spider lies dormant, its life processes at such low ebb that the small chamber of air proves adequate to its oxygen needs until the advent of warmer weather.

THE JUMPING SPIDERS

The line of the two-clawed vagrants culminates in the jumping spiders of the family *Salticidae* (see Plates 26–28; Plates XXIX, XXX). These are specialists. They stalk and attack insects with a precision and alertness not possible for myopic types. They are big-eyed experts that hunt during the daytime, and far out-shine the wolf spiders and their kin. Life in the sun seems to have produced a variety and brilliance of coloration not matched by any other spiders; a display of this ornamentation is part of their courtship ritual (see Chapter 5). Quite friendly little creatures, they sometimes sit upon a finger and follow one's every move with an attention not ordinarily manifest in arthropods bound by complex instinctive patterns. Fine eyesight has made them the outstanding spider extro-verts. The largest eyes of our spotted *Phidippus audax* (Plate 25) are capable of receiving a sharp image (perhaps ensuring recognition of another's species and sex) at a distance of ten or twelve inches. Awareness of moving objects by the four pairs of eyes, each of which receives different sized images, is possible at a much greater distance. The jumping spider spies its prey in the distance, creeps slowly forward until very near, then leaps suddenly upon it.

Almost all jumping spiders are small; few much exceed one-half an inch, and most fall far short of that length. The short, stout body, the rather short legs, and the distinctive eye arrangement make them one of the most easily recogniz-able of all groups. The rectangular cephalothorax is large and wide, squared off in front, and often quite high. As in the wolf spiders, the eyes are set in three distinct rows: four, two, and two. Those of the front row (small in wolf spiders) are greatly enlarged—the middle pair especially, which resemble large, smoky pools and, well supplied with rods, give the most perfect image. Above the front row is a second row of two tiny eyes, and behind these a third row of two larger ones. The abdomen is often oval, but may be thick and wide, or greatly elon-gated, to conform with the cephalothorax. Over the whole body is usually pres-ent a thick covering of colored hairs forming an even blanket, as well as longer hairs and spines that add special adornment according to the species.

The jumpers run, leap, and dance gracefully on legs of moderate length. The first pairs of legs are usually longer and thicker than the hind ones, especially in the males, whose front legs are in addition bedecked with conspicuous plumes and ornaments prominently displayed during courtship. It is a surprise to find the hind legs, which are most used in jumping, neither modified nor strengthened as they are in such animals as kangaroos and frogs. Apparently the small size and slight weight of the spiders make possible those tremendous leaps up to forty or more times the body length. They leap from stem to stem with ease and seeming abandon and are saved from falls by dragline threads laid down wherever they go. They have been observed to leap away from a building and catch insects in flight, a feat that demonstrates the remarkable coordination of their senses and their superiority among spiders as hunters.

The jumping spiders spin retreats of thick, white, slightly viscid silk in crev-ices, under stones on the ground, under bark, or on foliage and plants. Many retire to these little white bags at night and during cold days, and also use them as headquarters for molting and passing the winter as juvenile or hibernating adults. The females lay their eggs in the retreats, usually in spring or summer, and may be found guarding their young after the hatching. Often many retreats are found grouped close together under a single stone.

The salticids abound mainly in tropical regions. There live a bewildering number of different types, many of them glittering like gems in an infinite variety of patterns. Although less abundant in the United States, more than 300 different species occur and others are constantly being discovered. Many of these penetrate far into the north. Some live on the ground but most live on vegetation. A characteristic element of the leaf mold of the whole temperate zone is the tiny, smoky-gray species of *Neon*, whose greatly enlarged dorsal eyes shine out of a body only one-tenth of an inch long. R. W. G. Hingston, the British naturalist, found a plainly marked jumping spider 22,000 feet up on Mt. Everest, a height at which few animals of any kind can live. Tolerant in another way are the few species that have become domesticated. The graceful zebra spider, *Salticus scenicus* (Plate 26), hunts on fences and the walls of buildings, and is as common in America as in Europe.

A number of jumping spiders exhibit such an amazing resemblance to ants that they are called "antlike" spiders (Plate 26). In the beginning these spiders, probably through mere chance, gained a superficial resemblance to ants by developing slender, cylindrical bodies and quite long legs. (In the same way, other salticids became plump, and came to resemble certain flea beetles through the shortening of the body into a globular form.) It is only natural that these spiders should run over the soil or vegetation much as the ants do—sometimes even in association with the latter. There are profound differences in structure between the two, and even the most antlike of spiders does not bear too favorable a comparison with an ant when parts of the body are closely studied. However, it is an entirely different matter when the spiders are alive and moving. Then they exhibit such an exacting simulation of ants that they are able to deceive even trained naturalists—and to some of them the word "mimic" is applied with good reason. All the antlike spiders have quite slender bodies and relatively long and thin legs. In certain cases there are deep constrictions in the cephalothorax and abdomen, with these parts narrowed to expose the pedicel. In other varieties only the abdomen is constricted, or a seeming division of the body into several segments is accomplished by white bands across the abdomen and cephalothorax without actual physical constriction. The spiders are mostly small, and approximate in size and color the ants they reputedly mimic.

Whereas the physical resemblance to ants may be a natural consequence of exploratory body growth within normal range of the family, thus mere parallelism, there is reason to believe that some degree of physical immunity, perhaps only slight at first, was the result of the antlike form. To the best mimics would accrue the greatest immunity from those normal enemies of spiders that hesitate to attack ants. The imitation of ant movement has been instrumental in bringing even greater advantages to the spiders, and probably has made unnecessary more profound changes in the body itself. The mimics assume the particular stance, walk with the same gait, elevate the abdomen, and move their legs in the manner characteristic of the models. However, when these antlike spiders are disturbed, the assumed posturing and gait are usually dropped, and the spider departs the scene in spiderlike fashion. Whether the mimicry is a real thing or just the figment of the observer's imagination remains a moot point, but the advantages to the spider are unquestionable. Our best physical mimics belong to the genus *Synemosyna.* The commonest species is *Synemosyna formica* (Plate 26), a slender black or brownish spider about one-quarter of an inch long, with deep constrictions in the cephalothorax and abdomen. Amazingly antlike in form, it walks and runs much as do ants, and uses its front legs as ants use their antennae. Our *formica* does not run in ant columns or live in nests; it derives advantage only from its form. A Florida relative is golden brown in color and has been found running on foliage with ants of the genus *Pseudomyrma;* if this is its habitual environment, presumably it would have even greater immunity than its darker congener.

Although we have several other genera and species of antlike jumping spiders, mention will be made only of one more. *Peckhamia picata* (Plate 26) is a small black spider with lightly constricted abdomen and quite thick front legs, which is undeniably antlike in form and actions, but it not identified as the mimic of any particular ant. This spider does not walk in a straight line but, with abdomen twitching at intervals, "zigzags continually from side to side, exactly like an ant which is out in search of booty." The thicker front legs are used for walking and support of the forward part of the body, but the second pair is raised above the others and made to resemble the antennae of ants. The substitution of the second legs for the role played normally by the first pair suggests that the antlike gait and stance may have been acquired after the front legs had already been committed beyond redemption to another use. Even while feeding, *Peckhamia picata* "acts like an ant which is engaged in pulling some treasure-trove into pieces convenient for carrying," and keeps beating the prey "with her front legs, pulling it about in different directions, and all the time twitching her ant-like abdomen." A related species, *Peckhamia americana*, was observed by Prof. W. M. Wheeler running up and down the trees in Florida in files with the ant *Camponotus planatus.*

The Peckhams attributed the low fecundity of *Peckhamia picata* (said to produce only three eggs) to its resemblance to ants. As have most exponents of ant mimicry, they assumed that ants have few enemies and reasoned that protected mimics would not have to produce so many offspring to maintain their normal population. It seems to be true that many predators shun ants or find them distasteful and that antlike spiders profit from this aversion. However, insufficient data on ant mimic fecundity are available to warrant the conclusion that the real protection mimics enjoy results in lowered egg production. Instead, low production seems to be related to body size and minimum egg size, which limit the number of eggs that can be matured for a single laying. Likewise, immunity to the attack of Pompilid wasps often cited as evidence for ant mimicry is largely a consequence of size inasmuch as most antlike spiders are too small to serve as larval food for the wasps.

The ant mimicry of the jumping spiders is distinct from that of other spiders in that they never enter and live in the nests of ants. The dark of ants' nests would take from them full use of the sense that has brought them greatest success among the vagrants.

We owe to Dr. and Mrs. Peckham the most important studies of our typical salticids, which comprise a large fauna of several hundred species assigned to more than thirty genera. It must also be reported that their 1909 *Revision of the Attidae of North America*, now far outdated, is still the only comprehensive review of our rich fauna. During recent years the family has begun to receive the attention deserving of such a large, varied and handsome group, and much biological and systematic work is now in progress. Some of our representative species were noted by the Peckhams in the chapter on Courtship and Mating. In general it can be said that the salticids live such stereotyped lives that the habitus of the few cover the many.

Our largest salticids (often more than one-half an inch long) belong to the genus *Phidippus* (see Plates 27, 28), hairy jumpers with distinctive epigamic features on the carapace and legs of the males. Not so gaudily colored, the females are sometimes quite as handsome. Many species of *Phidippus* live all over our country and command attention by their size and alertness. A black species with white markings is widespread *Phidippus audax* (Plate 28), which often has white bands of hairs on the carapace and the black abdomen is marked by a large triangular white spot at the middle and behind two pairs of small white spots. Mostly black in the North, *audax* is much larger and striking clothed with white in the South. A larger relative found mainly in Florida, *Phidippus regius*, often has orange or reddish patches on its carapace and abdomen, and is one of the most variable members of the genus. Some other species, called red spiders by the Peckhams (*whitmani*, *cardinalis*, and *texanus* among many others), have the carapace and abdomen covered above with bright red hairs; in others the carapace is variously provided with bands or spots of white or yellow, pencils of black hairs, and snowy-white hairs. The first legs of males feature white, black, or mixed fringes of long hair, or even iridescent plates and enlargements. Even more numerous than our *Phidippus* are several scores of near relatives belonging to the closely allied genera *Eris* and *Metaphidippus*. Of these the most colorful species is *Eris aurantia* with green iridescent scales on the carapace and a brown abdomen gleaming with green, golden, and pink scales. This brilliant spider of tropical America also has a wide range in the United States from coast to coast, even as far north as Kansas and Illinois.

In this token treatment of our *Salticidae* we must omit even mention of some of our most interesting and exciting jumpers. Some of our *Habrocestum* are house spiders and often in buildings we find the tropicopolitan and other species here as immigrants. One of our charming and distinctive genera is *Pellenes*, small jumpers rarely one-quarter of an inch long with scores of species in North America. The females are a drab lot with grays, browns, and dull striping predominating. The little males, which hop around like tiny toads, have peculiar modifications of form, color, or ornamentation appearing mainly on the first and third legs. These features are put to use during the courtship dances.

THE CRAB SPIDERS

The superficial resemblance of some two-clawed vagrants to crabs has given them the name of "crab spiders," and their ability to move sidewise or backward with great facility enhances the pertinency of the appellation. For the most part, they have short, wide, considerably flattened bodies, and some of the legs are extended laterally at nearly right angles to the body. Those that most nearly resemble true crabs are various ambushing species with short, thick legs, the first two pairs of which are held sidewise and twisted somewhat off the normal axis so that the lateral surfaces become nearly dorsal in position. The laterigrade spiders were derived from typical hunters with normal prograde locomotion, and they exhibit varying degrees of development between extreme variation and near normality. The laterigrade form and attitude appear sporadically among other families of spiders, but those discussed in this section seem to form a single line.

The crab spiders wander about freely on the ground and on plants and have come to rely almost entirely on strategy and the chase to capture insects. They spin no capturing webs; they ordinarily settle down in one place only at the egg-laying period, when they produce large, lenticular egg bags hidden and guarded for long periods by the mother. Their flattened bodies fit them eminently for life in narrow crevices, under bark, or in debris, but many of them lie appressed to the surface of plants or on rocks or soil in the open. Some come out from their hiding places only at night, but others seem to be committed largely to the capture of day-flying insects. Their reliance on touch rather than sight would appear to make them equally expert hunters by night or by day, and many hunt at either time.

The large size of the laterigrade spiders of the families *Selenopidae* and *Sparassidae* (as compared with the relatively small typical crab spiders) has occasioned their title of "giant crab spiders." Only a dozen species of this tropical group are found within the borders of the United States, and these are limited largely to Florida and our southwestern states. All are one-half an inch or more in body size and have long legs of nearly equal length.

Amazing for their celerity are the extremely flat species of *Selenops* (Plate 23), which, closely pressed against rock surfaces in their southern Arizona canyon habitats, easily elude capture by whisking like squirrels into narrow crevices. These very flat vagrants are also common in houses and buildings and hide behind pictures on the wall and furniture. These domestic species are predators of insects that live in houses or manage to gain entrance during periods of heavy flights. While in Panama many years ago one kind lady showed the author several of these flat spiders stationed on the walls of her kitchen, knew the location of every one in her establishment, and praised them as being very beneficial.

Less agile are the plumper species of the family *Sparassidae* best represented by the tawny or brownish species of *Olios*, which are often discovered with their bulky egg sac. When menaced, they retreat into the spiny security of prickly pear or cholla cacti. Our largest species is *Olios fasciculatus* (Plate XXXII), which wanders into houses and buildings to bring some consternation to the housewife. This harmless spider achieved sudden newspaper fame as the "barking spider," in some communities of west Texas, and this despite the fact that it is mute and not provided with sound-making organs of any kind.

The best known giant crab spider is *Heteropoda venatoria* (Plate 18), a huntsman spider, a species that occurs around the world in the tropical zones, and penetrates northward into Florida, where it is quite common, and into the subtropical regions of Texas. This spider came originally from the Asiatic mainland, where many close relatives live and from which locality we have acquired many of our commonest domestic insects. It was the belief of the Reverend Henry McCook that the huntsman was distributed by means of ballooning threads, and that its tropicopolitan distribution was determined by the prevailing trade winds. This may be partially true, since it is known that flying spiders cover tremendous distances, but its prevalence can be attributed also to a great climatic tolerance and domestic habits that make it an ideal emigrant in goods carried by boat. None of the close relatives of the huntsman spider has been disseminated in a like manner, even though the habits of all the species are similar and the opportunity of transfer was present almost equally to all of them.

Often having tawny bodies over an inch long and outspread legs spanning three or more inches, *Heteropoda venatoria* is the commonest and best known of tropical house spiders. They are generally welcome because of their depredations on roaches and other disagreeable insects that abound in the poorly constructed homes in the tropics. Although they live under bark and in similar situations in the open, they show a marked preference for houses, barns, docks, sheds, and other buildings of man. Hidden away in crevices by day, they come out at night and hunt on the walls. Because they are frequently carried into northern regions in banana bunches, these large spiders are often called "banana spider," a name which fits equally as well many large ctenids and tarantulas.

The typical crab spiders are placed in two families, the *Philodromidae* and the *Thomisidae;* these are small spiders that rarely exceed one-third of an inch in body length. More than a hundred species of each family live in the United States and they are encountered as commonly in the North as in the South. In the philodromid line of these crab spiders we find these characteristics: elongate body, legs quite long and all about the same length, brushes of hairs on the legs, and a pair of adhesive claw tufts on the tip of each tarsus. The philodromids are swift runners, and move easily on precipitous surfaces. For the most part they live on vegetation and they are colored to conform rather closely with the particular surface on which they sit. Especially well camouflaged are the running crab spiders of the genus *Philodromus*, which, with long legs spreading far sideways, press flat against the surface of a tree or stem. The common *Philodromus pernix* of the northeastern states, and its many close relatives all over our country, have mottled bodies that closely resemble the bark of trees; and those of domestic inclination are not easily discerned against the weathered boards of fences and buildings. Other species are more brightly colored, and, as is the case of the widespread *Philodromus rufus*, prefer the colored leaves of various bushes and trees, under which they attach their tiny egg sacs. One of the commonest western representatives of this genus, *Philodromus histrio* (better known as *virescens*), has the same bluish gray color as the sagebrush on which it is most often encountered. Other philodromids run on the ground, where they hide in grass or plants. Our species of *Thanatus* (Plate 23) can climb well, but they often hide under stones and behave like running ground spiders. Our numerous species of *Ebo* are usually swept from vegetation but some live on the ground as well. A tiny species, *Ebo pepinensis*, common on the open sand along the edges of streams and lakes in southern Minnesota, matches the sand almost exactly in color; it remains unnoticed until accidentally disturbed, whereupon it runs a few inches and again lies perfectly still. The greatly elongated species of *Tibellus* (Plate 23), straw-colored and lightly marked with dark, narrow lines, frequent the grasses in meadows, lying parallel to and close against the stem. Easily visible when moving, these spiders will stop suddenly and appear to vanish from sight in their natural environment—characteristics which they have in common with the majority of philodromids.

The typical crab spiders of the family *Thomisidae* have become specialists in ambush; they accomplish by surprise what the jumping spider is able to achieve through superlative eyesight and stealthy approach. Fortified with extremely potent venom in compensation for weak chelicerae, the small crab spiders are formidable creatures that will attack insects and other spiders much larger than themselves.   The thomisids were probably derived from types similar to the philodromids but their specialization has markedly changed them, making them into phlegmatic creatures that excel as ambushers.   Their short, wide bodies are supported by legs of very unequal size, the first two pairs being quite long and robust and the hind pairs considerably shorter and weaker; their habitus is notably crablike.   They have sacrificed ease of movement for a leisurely life in flower heads or on the ground, and have lost the brushes of hairs beneath their legs and the tarsal claw tufts present in their forebears.

The species of *Ozyptila* and *Xysticus* (Plate 23) are preeminently spiders of the ground; their colors, dull grays, browns, and blacks, mingle with the leaves and organic debris of the soil.   They squeeze their flat bodies under bark and into cracks.   The mottled, greatly flattened species of *Coriarachne* simulate to a remarkable degree the bark of trees or the old wood of fences and houses on which they hunt.

The ambushing crab spiders that live on vegetation and in flowers are much more brightly colored than the ground forms, but these tend equally to cryptic coloration.   The delicate green *Synema viridans*, for example, lives on foliage, while some of the whitish or more colorful species of *Xysticus* are distinctly flower or foliage forms.   The best known flower spiders of the North Temperate Zone are the numerous species assigned to three closely allied, often confused genera, *Misumena*, *Misumenops*, and *Misumenoides* (Plate 24): handsome white, yellowish, or saffron-yellow spiders often marked with black or red bands and spots.   All are ambushers, and obtain their livelihood by strategy.   They are usually found in the heads of flowers; there, simulating the phlegmatic assassin bugs, they lie motionless in wait for insects seeking pollen or honey.   Large and seemingly dangerous bees and wasps, large-winged butterflies and moths, and a host of winged insects are seized and quickly dispatched by the pygmy ambusher.

In keeping with their habit of deception, these ambushers are known to change color from white to yellow, to conform with the substratum of their hunting ground.   In this connection it should be noted that while they may be found on a variety of colored flowers, a very high percentage occur on white or yellow ones.   In the fall A. S. Pearse, the American ecologist, found that 84 percent of all the white spiders (perhaps two or three species) were on white flowers, and 85 percent of the yellow spiders were on yellow flowers.   Only 6 to 10 percent of the spiders were found on flowers other than white or yellow.   Being homochromous with their flower station seems to bring them some advantage in their hunting as well as a measure of immunity from their enemies.   It is well known that flying insects avoid light-colored flowers in which sit dark spiders or insects, or small dark objects placed there by investigators.

The ability of *Misumena vatia* (or *calycina*) (Plate 24) to change its color from white to yellow and vice versa was first noted about seventy years ago.

This fact engaged the attention of many naturalists and led, in some instances, to erroneous application of the same principle to other spiders on little evidence, to fantastic claims of change through many hues that have no basis in fact. It can easily be demonstrated, however, that *Misumena vatia* and many of her cousins can change, in the course of a week or more, from white to yellow on a yellow flower or an artificial yellow substratum. The action is reversible, usually requiring only five or six days. There is reason to believe that the immature stages of this spider are always white, and that the changes in color are possible only for mature females, as was claimed by Eugen Gabritschevsky, a French biologist. However, both juvenile and adult examples of the allied *Misumenoides aleatorius* of the United States may be shining yellow, and are reputedly capable of changing back to white.

Because of their peculiar body forms, certain crab spiders have been singled out as receiving some sort of protection from their natural enemies through their resemblance to inanimate objects. *Phrynarachne rugosa* of Africa and Madagascar is said to resemble in form and color the fruit of a common tree in its forest home. Another species of the same genus, *Phrynarachne decipiens* of Malasia (once assigned to the appropriate name of *Ornithoscatoides*), is reputed to resemble the excreta of a bird, and the illusion is complete when the spider has fashioned its characteristic web. Other thomisids have been compared to dried seeds, leaf buds, and various flower parts.

THE RUNNING SPIDERS

The running spiders are two-clawed vagrants that wander about over the soil and on vegetation, aided in their movements by adhesive tarsal claw tufts. In almost all instances the front legs are directed forward and locomotion is normal or prograde, as contrasted with the laterigrade maneuvering of the crab spiders. Their bodies, usually elongated and often subcylindrical, are furnished with quite stout legs, which propel them with great speed. Some rarely leave ground hiding places under stones and debris, while others climb actively over vegetation and make their retreats in plants. These running hunters, probably typical of the prototypes from which have separately arisen the crab and jumping spiders, have become specialists in their own way. Whereas they must concede superiority in daylight hunting to the jumping and wolf spiders, and also to a few of the ambushing crab spiders, they have less competition at night. Their distrust of sunlight has kept them essentially night hunters—or shy shade hunters under debris by day—and they have as chief competitors the phlegmatic crab spiders and nocturnal wolf spiders.

The eyes of the running spiders are for the most part set close together in a small group near the front of the head, are of small and essentially equal size, and they are not placed strategically for sight-hunting. They probably see moving objects, and may have fair close-range vision, but sight does not appear to play much of a part in their nocturnal foraging. By day they remain as far as possible in the shade of litter; they run rapidly across sunny open spaces until able to hide themselves, when they again resume a more deliberate pace. Their front legs searchingly test the terrain, and they are uncertain of the character of objects—even of the near presence of a prospective mate—until they actually touch them. However, they move about with a seeming boldness that belies this, and they can hold their own with longsighted spiders when they come to grips.

Many running spiders make flattened, tubelike retreats (Plate XXXI) of white silk, in which they remain by day and in which they molt, mate, and deposit their eggs. The ground-loving types place the eggs under stones or in dark recesses under debris. The plant spiders bend leaves or fold blades of grass, then bind them down with silk to provide cozy domiciles. The eggs, held in the usual two sheets forming a lenticular sac, are guarded by the mother; she often remains with the young until they are hatched and dispersed. Some ground species cover the eggs with debris or camouflage them in other ways before leaving them to their fate.

The vagrants described in this section constitute a related assemblage of seven to ten families of which several hundred species occur within our borders, but passing mention can be made of only a few.

The vagabonds of the family *Gnaphosidae* are mostly ground spiders of somber coloration with few contrasting markings; the dull grays, browns, and blacks deriving from a covering of short hairs that gives them a velvety appearance. More flattened than their near relatives, the clubionids, they differ from the latter also in having the anterior lateral spinnerets widely separated. Typical of the group is *Herpyllus ecclesiasticus* (formerly *vasifer*) (Plate 31), the "parson spider," a blackish species one-third of an inch long with bright white markings on the abdomen, which lives outside under stones, but is even more common on the walls and floors of houses. A near relative is *Scotophaeus blackwalli*, the "mouse spider" of England with similar domestic habits, now an emigrant into this country common along the California coast and rapidly widening its range. Some of the gnaphosids, and notably bold, powerful *Drassodes*, commonest in the North and often an inch long, trail a band of silk that serves to entangle the legs of opponents while they spar and grapple at close range.

Some of the gnaphosids are more brightly colored. Outstanding are the species of *Poecilochroa* (*Sergiolus*) (Plate 31), some of which have a bright orange cephalothorax and a black abdomen pleasingly variegated with white or colored stripes and spots. The shining, coal-black species of *Zelotes* (Plate 31) run over the soil in company with their close relatives, the small brownish, tawny, or gray *Drassyllus*. They hide under stones or leaves, and often attach beneath stones

Plate 17

M. W. Tyler

a. *Mastophora bisaccata* with casting line

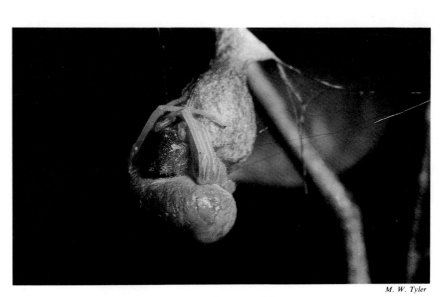

M. W. Tyler

b. *Mastophora bisaccata* with egg sac

BOLAS SPIDER

Plate 18

a. Huntsman spider, *Heteropoda venatoria* on light switch

a. Southwestern giant crab spider, male *Olios fasciculatus*

GIANT CRAB SPIDERS

Plate 19

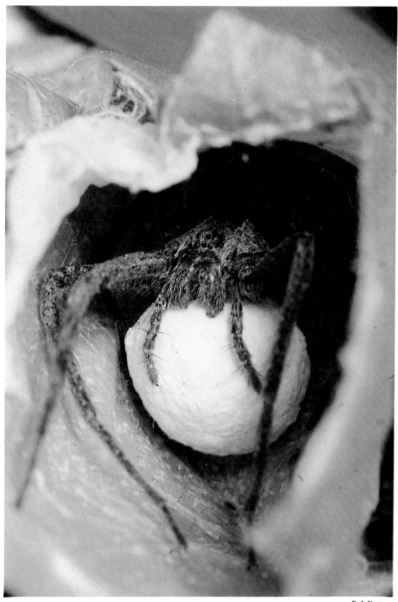

B. J. Kaston

Pisaurina mira WITH EGG SAC

Plate 20

a. *Dolomedes albineus*, female

b. *Dolomedes striatus*, female with egg sac

c. *Dolomedes triton*, male

d. *Dolomedes triton*, female

FISHER SPIDERS

Plate 21

H. K. Wallace

a. *Lycosa punctulata*, female

H. K. Wallace

b. *Geolycosa rogersi*, female

H. K. Wallace

c. *Geolycosa micanopy*, female

H. K. Wallace

d. *Lycosa carolinensis*, female

WOLF SPIDERS

Plate 22

GREEN LYNX SPIDER, *Peucetia viridans*, WITH EGG SAC

Plate 23

H. K. Wallace

H. K. Wallace

a. *Thanatus*, female

b. *Tibellus duttoni*, female

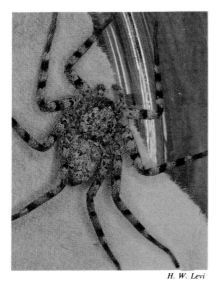

H. W. Levi

H. K. Wallace

c. *Selenops*, female

d. *Xysticus*, female with egg sac

CRAB SPIDERS

Plate 24

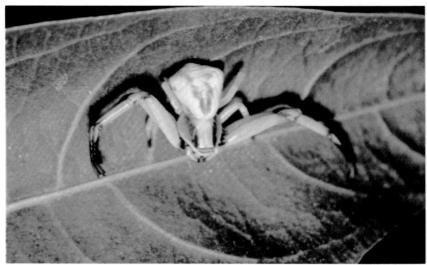

Bucky Reeves

a. *Misumenoides aleatorius* on leaf

H. K. Wallace

b. *Misumenoides aleatorius*, male

H. K. Wallace

c. *Misumena vatia*, female

CRAB SPIDERS

Plate 25

B. J. Kaston

a. *Phidippus audax*, female

H. K. Wallace

b. *Lycosa carolinensis*, male

FACE VIEWS OF JUMPING AND WOLF SPIDERS

Plate 26

H. K. Wallace

a. *Synemosyna formica*, female

H. K. Wallace

b. *Peckhamia picata*, female

H. K. Wallace

c. *Icius elegans*, male and female

B. J. Kaston

d. *Salticus scenicus*, female

ANTLIKE JUMPING SPIDERS

Plate 27

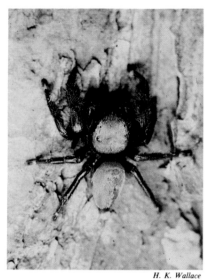

a. *Phidippus apacheanus*, male

b. *Phidippus apacheanus*, female

c. *Phidippus clarus*, male

d. *Phidippus* from Florida, female

JUMPING SPIDERS

Plate 28

a. *Phidippus audax*, male

b. *Phidippus otiosus*, female

c. *Phidippus variegatus*, female, reddish phase

d. *Phidippus variegatus* female, gray phase

JUMPING SPIDERS

Plate 29

a. *Marpissa bina,* female and male

b. *Marpissa grata,* male

c. *Marpissa pikei,* male

d. *Marpissa pikei,* female

JUMPING SPIDERS

Plate 30

a. *Castianeira descripta*, female

b. *Castianeira variata*, female

c. *Castianeira trilineata*, female

d. *Clubiona obesa*, female

RUNNING SPIDERS

Plate 31

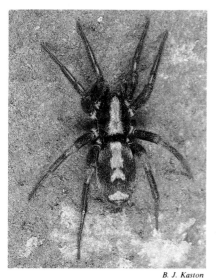

a. *Herpyllus ecclesiasticus*, female

b. *Poecilochroa variegata*, female

c. *Anyphaena pectorosa*, female

d. *Zelotes*, female

RUNNING SPIDERS

Plate 32

a. *Loxosceles reclusa,* female

b. *Loxosceles reclusa,* female, enlarged

c. Spitting spider, *Scytodes fusca,* female

d. *Dysdera crocata,* female

PRIMITIVE SPIDERS

their tough pinkish or brown egg sacs, variously covered with debris or lacquered with saliva and excrement to form a horny covering as a deterrent to penetration by predators. The many genera in our fauna, including *Gnaphosa*, *Callilepis*, *Haplodrassus*, *Orodrassus*, *Nodocion*, and others, have representatives in all parts of our country. Some of our species also occur in Europe and many others are most closely allied to those in that part of Holarctica. Smallest of all gnaphosids are the exceedingly active species of *Micaria;* as many as fifty or more species of different color, pattern, and size make up our rich fauna.

An extremely rare relative of such ground spiders is *Prodidomus rufus* (family *Prodidomidae*), a pink-bodied spider occasionally found in houses. It has been taken on Long Island, in a few places in the South, and recently in southern California.

The vagrants of the families *Clubionidae*, *Anyphaenidae*, and others not as yet fully established are less flattened than the gnaphosids, often have longer legs, and have the fore spinnerets set close together. Those that live on plants have well-developed claw tufts and are good climbers. Mostly whitish or brownish, one-fourth of an inch long or smaller, and represented by numerous species in the genera *Clubiona* (Plate 30), *Chiracanthium* (Plate XXXI), *Anyphaena* (Plate 31), and *Aysha*, the plant hunters live in flat tubular nests, open at both ends, in rolled leaves or under bark. Some of these also run over plant debris and nest under stones.

The clubionids and their near neighbors that habitually run on the soil exhibit far more diversity in size, appearance, and coloration than do the conservative plant forms. Among the largest are the inch-long, speckled, grayish tramps (*Syspira*) that wander over the sand of our southwestern deserts and resemble the wolf spiders. Our reddish *Liocranoides* (or *Titiotus*) often of similar size or even larger, known from a single species in the Appalachian Mountains and by many more in California, favors the caves and deep rock fissures in the canyons and along the streams of their respective regions. Intermediate in size are many gaudily colored *Castianeira* (Plate 30)—golden, bright red, or black with stripes and spots of red, yellow, or white—which resemble some of the wingless mutillid wasps. And smallest of all are the gray, red, and black species of *Scotinella* that run with great speed.

In this series of ground hunters are some that compare with the best of the ant mimics, and still others that have become intimately associated with ants and live in their nests. Whereas the jumping spiders are keen-sighted, diurnal types whose mimicry is influenced by reliance on sight, the shortsighted, chiefly nocturnal clubionids have developed in a different pattern. Some emulate the jumping spiders by running about during the day—even in the hottest sun—in open places, frequently in company with ants whom they resemble in size and color. Many exotic types are almost exact physical mimics. Our numerous species of

Micaria, as well as the smaller forms of *Castianeira*, are sun spiders. Their slender golden or black bodies, constricted or crossed with white bars, are covered with flattened and iridescent or brightly colored scales. Moving actively about with quivering front legs, these beautiful vagrants are less susceptible to attack than are other species, and are actually shunned by certain insect and spider predators in the same way as are true ants.

A similar immunity is probably enjoyed by our many antlike species of *Scotinella* and *Phrurotimpus*. Although not usually diurnal, they are often associated with ants in the soil debris, and occasionally are seen running with them during the day. A few live as myrmecophiles in close association with or even in the interior of the ant nest itself. While little can be said of their life underground, they seem to be tolerated by the ants, and it is known that they feed upon ant pupae and small insects living in the nest. Doubtless they live in comparative safety in their little silken cells, in this way isolated from the ant colony even when in their midst. These myrmecophiles agree in appearance and color with the particular ant in whose nest they live. *Scotinella formica*, a black species with a shining sclerotized plate on the abdomen, frequently is found in the nests of the black ant, *Cremastogaster lineola*, in some of our eastern states. The rare *Scotinella britcheri*, a yellowish spider of New England, lives with yellow ants, and shows the effects of its underground existence by lacking much of the normal dark pigmentation around the eyes.

Before leaving this series of tramps and vagrants, mention should be made of some relatives of the *Clubionidae* that barely reach our southeastern states and western states bordering Mexico. The wanderers of the family *Ctenidae* are often giants that climb over foliage at night or run over the detritus on the ground. Some resemble the wolf spiders in appearance and are covered with dense coats of tawny hair. Our several native species live in our Gulf states and Florida and in the lower Rio Grande Valley of Texas. Many large ctenids are brought into this country on banana bunches but this traffic has declined with refrigeration and more efficient fumigation of banana boats. Among them have been large species of *Cupiennius*, *Ctenus* (see Plate XXXII), and *Phoneutria*, most of them probably from Central American ports. Dr. Wolfgang Bücherl, of the Instituto Butantan, São Paulo, Brazil, reported that the most active spider venom known was that of *Phoneutria*, especially the species *fera*, whose paralyzing neurotoxic venom marked it as the most dangerous species in South America.

The family *Zodariidae* is represented in the United States by several species of *Lutica* that live in the sand dunes behind the cool beaches of southern California and on the Channel Islands just off the coast. *Lutica* is best identified by the presence of long, robust anterior spinnerets and the reduction of the median and posterior pairs to mere vestiges. These sand spiders live in the small dunes tied down by binding plants well back from the high tide lines. They are nocturnal and come to the surface at night to hunt various beetles and other insects that drop on the sand. When disturbed, they dig actively into the fine sand and are quickly lost to sight; presumably they spin a loose, tubular retreat deep in the cool and sometimes moist sand and stay there. These whitish spiders, about one-half an inch in length, have their nearest relatives in North Africa, where similar species of the genus *Laches* live under stones in the hot deserts.

The adaptive radiation of our species of *Lutica* is reflected in the history of the Channel Island archipelago, consisting of eight islands which fan out into the Pacific Ocean from Los Angeles as a center, and which are thought to be the remnant of an old land mass connected on one or more occasions with the mainland. It is postulated that a single stem species of *Lutica* was widely distributed and had access by land to even such remote islands as San Nicolas and San Clemente. The submergence of the old land mass called Catalinia left several isolated populations of *Lutica* on the high points, now the Channel Islands. These sand zodariids of retiring nocturnal habit now are restricted to confined dune habitats and do not balloon or disperse from them as do many other spiders. They have responded to this isolation since Pleistocene times by producing several distinctive, allopatric populations. Those from the two most distant islands, dubbed *nicolasia* and *clementea*, thought to have been isolated from the mainland for a longer period than the northern islands, are claimed to be valid species. Studies on these derivative sand zodariids are still going on.

Another zodariid, *Storena americana*, attributed to our fauna by George Marx on the basis of a specimen from Cohuta Springs, Georgia, more than likely was imported from the Bahama Islands. Many years ago I found many males and females of *americana* on South Bimini. They lived in the sand well back from the beach and had much the same habits as our *Lutica*. They dug formal tunnels into the sand, the walls of which were covered by silk, and closed them with wafer-type trap doors. When insects dropped on the sand, attracted by our Coleman lanterns, they opened their burrows and came out to capture their prey. Species similar to our *americana* are found in the Australian region.

In our extreme southwestern states and adjacent Mexico lives a small group of enigmatic spiders given special status as the family *Homalonychidae*. Their vagrant status is signified by possession of two smooth tarsal claws accompanied by claw tufts and a pair of spurious claws. The few known species are quite flat and live under stones in arid regions. Much still remains to be learned about their habits.

THE FUNNEL-WEB SPIDERS

The most generalized members of the three-clawed hunting spiders are the numerous, diverse representatives of the family *Agelenidae:* the funnel-web spiders.

It is in this group that we find close correspondence in basic features with members of various cribellate families discussed in Chapter 8. Some of our agelenids seem to be ecribellate amaurobiids; our typical hahniids seem to be ecribellate dictynids; some clubionids seem to be ecribellate tengellids; and other such pairs are numerous. The thesis that all spiders were once cribellate and that most eventually discarded this accessory spinning field to live freer lives, voiced by many older students, has now become a conviction of an increasing number of modern araneologists. This thesis postulates that our most derivative hunters, the wolves, lynxes, fishers, and even our jumpers were once cribellate. Further, it suggests that this divestiture is still going on, that in some families we have both cribellate and ecribellate members, and that such double identity may even be found in members of a single genus. Fortunately, in this chapter on hunting spiders we need not concern ourselves with the complex new family concepts now being proposed (these are still in a state of flux) and can maintain for the most part the conventional nomenclature.

Only brief mention can be made in this short section of the impressive number of species of *Agelenidae*, exceeding 400, that occur in our country. Most areas have a representative quota of genera and species but the far western region from Baja California to and above British Columbia—the Californian region—is blessed with a great number of endemic genera and scores of species, many of the latter still undescribed. Responsible for this plethora of taxa, conceived in a changeable geological history, are the many circumscribed habitats resulting from the height and position of the great coastal and inner mountain ranges, separated by hot arid valleys, and further partitioned by deep canyons to form many ecological enclaves. The adaptive radiation occasioned by this region has also been operative in many other plastic spider families.

In physical appearance the typical agelenid is far less changed from the hypothetical prototype of the group than are the wolves and fishers and jumpers. Unlike these big-eyed hunters, most agelenids are shy and hide under debris or in vegetation in their funnel webs. Their hunting activities are mostly contained within the limits of the silken sheet that they lay out over the terrain. Since they spin a web and use it to trap insects, they are often called sedentary spiders, but they represent a separate line from the aerial sedentary spiders.

The cephalothorax of the agelenids is suboval and convex and the eyes typically lie in two rows near the front edge of the carapace. The eyes are not notable for size in any of the groups; they are far inferior to those of relatives who place reliance on sight when hunting. The legs are long and thin, particularly in the most active members, and are never provided with tarsal claw tufts to facilitate climbing. It can be said that evolution has caused the agelenids to modify their bodies, especially their eyes and legs, far less than their vagrant cousins have done. They are called generalized; but at the same time, in regard to their way of life they have become highly specialized.

The funnel web of the agelenids is little changed from the silken cell of their forebears and they still hide it under stones or logs, in crevices, and in deep vegetation. The funnel is open both in front and behind, thus providing the spider with a rear exit if it is menaced. From the outer opening of the retreat is spread an expansive field of white webbing—the sheet web—which may be placed on or just above the soil or suspended high in vegetation like a hammock. It forms a smooth runway on which flying insects can alight under the mistaken impression that is it a suitable landing field. Once down, they find it a spongy, yielding trap into which they sink and over which they drag their bodies with difficulty. For the spider, the sheet is a racing course; it is able to run over the surface with great speed in an upright position and catch the insect before it can reach the edge of the snare.

The typical agelenid sheet web, composed entirely of dry silk, grows up with the spider to adulthood, changing from a small, thin mesh into a thick blanket of considerable expanse. It grows by accretion, the result of incessant spinning by the active spider during most of its life. Upon a framework of long dragline threads, coming from the front spinnerets, and outlining the sheet, are put down many finer lines drawn from the hind spinnerets, which are moved from side to side like brushes and lay multiple filaments. The sheet web is rarely a simple, two-dimensional structure; suspended above it is a network of lines, placed in irregular fashion and attached to adjacent grass or twigs, which serves as a barrier to jumping and flying insects and often causes them to drop upon the sheet.

Our best known agelenids, the grass spiders of the genus *Agelenopsis* (Plate XV) and several related genera, scatter their sheets in immense profusion—over grass and shrubs, often high above the ground and frequently on and even inside houses. In the autumn these webs reach their greatest size and, seen in the early morning when they are covered with dew, carpet many acres of grassland. The spiders themselves are of moderate size, with bodies ranging from one-half to more than an inch long, and are colored variously with pale yellow to dark brown. The cephalothorax has light median and lateral stripes above, while a broader, speckled band runs the length of the abdomen. Three or four well-marked species, differing in size, color pattern, and habits, will be found in almost any single locality in the United States.

Agelenopsis sits in the mouth of her funnel retreat facing the sheet and poised for a swift foray out over its surface. A scarabid beetle drones through the air and strikes the aerial network; its flight is abruptly arrested, its heavy body plummets to the surface of the sheet; the lurking spider races swiftly and surely to the site of the disturbance. *Agelenopsis* wastes little time on small insects, seizing them quickly, but bulkier prey is approached with more caution. She rushes in to deliver quick bites, then retreats until the weakened insect can be approached and dragged into the funnel retreat for feeding. Many kinds of insects come her way, but the abundant grasshopper population of the grassland probably provides her with the highest percentage of her food.

The female grass spider lays her eggs in the fall and dies sometime thereafter, perhaps with the first hard frosts, her whole life spanning only a single year. The egg sac is a lens-shaped packet composed of two circular valves sewed together around the edges and is similar in form to those of the crab spiders and many others. Several sacs may be made close together. All are hidden in secluded places, frequently under loose bark on trees or lying on the ground. The sacs are fastened closely to the substratum and covered with silk in which bits of bark and debris are distributed, but this stratagem does not deter parasitic insects from laying their eggs in the masses. Investigation of the sacs even in the late autumn when the female is still very much alive and should be able to protect her young, will often disclose that the contents have already provided food for parasites that now occupy the cradle.

The feature that most easily identified our multitudinous array of grass spiders is the pronounced backward curvature (called procurvature) of both rows of eyes. Almost all other agelenids have these rows almost straight. Only some of our genera, notably *Agelenopsis,* have the apical segments of their posterior spinnerets much longer than the basal segments. The region west of the Rocky Mountains is especially well provided with grass spiders. About 100 species assigned to such expressive generic names as *Calilena, Hololena, Rualena,* and so on live in the ecological enclaves of that region.

Relatives, in size at least, of the grass spiders are the several species of *Tegenaria* found in the United States. One of these, the cellar spider *Tegenaria domestica,* is widely distributed in temperate and tropical parts of the world and is common in houses in our northeastern states. The web of the cellar spider is the same cobweb used so extensively many years ago by European and American peasants to staunch the flow of blood. Like *domestica,* several other *Tegenaria* have been brought into this country by trade and are rapidly widening their ranges. Our single native species, *chiricahuae,* lives in caves and other dark situations in Arizona and New Mexico.

Simulating the *Tegenaria* in superficial appearance are about twenty-five long-legged *Calymmaria,* that to some extent replace these imported species in California. Two species live in the southern Appalachian Mountains and are disjunct reminders that the genus was once more widely distributed. Our *Calymmaria* live in ground cavities in moist oak and coniferous habitats and spread out their sheet webs from such stations. A triad of species often occurs close together, as exemplified by one in a ten-foot-wide wooded canyon near Berkeley. "The largest species occurs among large rocks and logs at the bottom of the canyon, a medium-sized species occurs on the sides of the bank, and a third and smaller species occurs among leaves on the ground."

More than fifty species of *Cicurina* occur in North America and they are widely distributed from southern Canada south to southern Mexico. They are mostly shy, pale spiders that live in moist places under stones and ground debris. This genus of plastic spiders has responded in a remarkable way to environmental conditions and has produced eight-eyed, six-eyed, and completely eyeless species in the several species groups. Completely blind species are found in some caves of the southern Appalachians, in many of those in the Edwards Plateau of middle Texas, and in some of the states of Mexico. Two other genera of pale, lucifugous agelenids, *Blabomma* and *Yorima*, represented by about forty species, are restricted to the Californian region. They are usually found in deep ground detritus and these also have eight-eyed, six-eyed, and eyeless representatives, the latter from caves in Calaveras County, California. Vincent Roth found an almost direct correlation between reduction in body size, coloration, and eye size, in numbers of leg spines, trichobothria, and setae on the colulus, as he dug deeper into the leafmold, the smallest species being in the lowest layers.

The agelenides of the subfamily *Cybaeinae* (now regarded by some students as a distinct family *Cybaeidae*) have had a remarkable development in the Californian region and also have two or three disjunct members in the Appalachian Mountains. About fifty species are described with perhaps as many more still unnamed and they are placed in four quite similar genera. The many species of *Cybaeus* are usually found in damp forests under ground objects. Their inconspicuous sheet web consists of a "tube or flat panel several times as wide as the spiders attached by several strands of silk to adjacent supports." *Cybaeozyga* is notable mainly because of three blind, undescribed species living in the caves of Shasta County, California.

The hahniids of the family *Hahniidae*, easily recognized by having their six spinnerets in a transverse row, are near relatives of the agelenids. The species of *Neoantistea* spin small sheet webs, about two inches in diameter and lacking tube retreats, which are often seen in profusion covering small depressions in the ground. Since representatives of *Hahnia* and *Antistea* are usually taken by sifting ground detritus, in which *Hahnia cinerea* and its congeners often live in great numbers, nothing is known about their webs even though it is unlikely that they do not spin them. The females attach small, lenticular egg sacs to some surface and few eggs are enclosed. The small hahniids, the largest rarely as much as three-sixteenths of an inch in length, have a broad tracheal spiracle placed far in front of the spinnerets. An unusual feature of some hahniids is a stridulatory device, consisting of two patches of appressed setae lying above and around the petiole of the abdomen and a pick of short setae on the posterior surface of the carapace, but nothing is known about the sound presumably possible from this file and pick apparatus. An excellent revision of the three genera and twenty-eight known species of the Nearctic region has been published by Brent Opell and Joseph Beatty.

THE PRIMITIVE HUNTERS AND WEAVERS

There are various spiders whose features mark them as representative of the ancestral stocks from which the higher hunting types on the one hand, and the aerial web spinners on the other, are thought to have sprung. Some of these primitives are active vagrants that compete with the running ground spiders whom they resemble in general appearance and action. Others are wanderers that stalk over the terrain in deliberate fashion, groping with their front legs as they hunt. Most of them retire to some sort of base during the day, a silken tube or a padded corner, but few use silk with proficiency or place much reliance on it as a means of capturing prey (the curious warning threads of *Ariadna* and the tangled maze of *Diguetia* are exceptional web types). All of them are short-sighted, and because they are active mostly at night, sight plays a small role in their hunting. Almost without exception they are six-eyed—the anterior median pair having been lost very early in their history—and the eyes, usually placed far forward, are not notable in size. In two families, the *Plectreuridae* and *Caponiidae*, all eight eyes may be present; at the other end of the scale, most caponiids have lost all but the anterior median pair.

The most obvious generalized feature of the whole group is the retention of reproductive organs that are but little advanced beyond those of the mygalomorph spiders. The genital bulb of the male palpus is usually a simple vessel drawn out to a point, much like a syringe, and there is scarcely any development of the accessory processes so characteristic of higher spiders. The epigynum of the female is likewise unspecialized, and the orifices and receptacles that receive the emboli are still hidden beneath the genital groove. As far as is known, these spiders still use the primitive embrace of the tarantulas during the pairing, and in most of them both palpi are inserted simultaneously into the epigynum.

One group of these primitive hunters is made up almost entirely of active runners whose chelicerae are not soldered together at the base and instead are capable of separate movement. Present near the base of the abdomen are four usually distinct spiracles, the posterior pair opening into tracheal tubes. The unpaired claws may or may not be present on the tarsi. Most of these spiders have elongated bodies, and have legs of moderate length. All four of the commonly recognized families have representatives within the United States, but few species can be mentioned.

The only member of the family Dysderidae found within our borders is the cosmopolitan *Dysdera crocata* (Plate 32), a one-half-inch long, orange-brown species with a gray abdomen, probably introduced into this country from the Mediterranean region, where many other genera and species occur. In our country *crocata* is rarely found in natural areas, in fact it occurs only sporadically over most of it, and prefers situations near buildings, where it hides under stones, boards, and other litter. An oval cell of closely woven silk serves as a retreat; within it are placed the eggs without a special covering or egg sac.

Quite similar in appearance to the dysderids are the small vagrants of the family *Caponiidae*. One of their unusual features is the absence of book lungs; all their respiratory spiracles open into tracheal tubes, a condition paralleled only in some of the aerial spiders. Most caponiids have two eyes, the anterior median pair, which are quite large and dark in color. Reminiscent of their earlier condition, however, is the genus *Caponia* with all eight eyes present and in *Nopsides* of Sonora and Baja California a four-eyed transitional condition is found. Except for a few species in southern Africa, the caponiids are found only in North and South America. The several species that occur in our southwestern states belong to the genera *Orthonops* and *Tarsonops* and these have the metatarsi of the legs provided with translucent keels along the ventral line and rounded apophyses at the base. These caponiids spin a silken cell beneath rocks in arid situations and then come out at night to hunt. Their food habits are little known but they have been seen feeding on aerial spiders and are sometimes found in the webs of social spiders. An undescribed caponiid from central California has eight eyes and lives in the duff in the redwood country.

The six-eyed vagrants of the family *Oonopidae*, most of them less than one-sixth of an inch long, live in leafmold or under stones, where they feed upon tiny animals ignored by larger spiders. Many are colored bright orange and have hard plates on the abdomen; others are whitish or pale yellow, with softer abdomens. These pretty spiders, of which only about twenty species are known to occur in our southern and southwestern states, run rapidly when they are disturbed. One of the smallest is *Orchestina saltitans*, a midget about one-twentieth of an inch long with a soft abdomen. It penetrates quite far into the northeastern states where it lives for the most part indoors as a domestic spider. It may occasionally be seen hanging by its threads from a lampshade, foraging in the midicine cabinet, or running among books on desks. *Orchestina moaba* of our Southwest lives in leafmold in arid foothill country. Several pale species of *Oonops* occur from California and Arizona across the southern states into Florida. Along with these are a number of bright orange species of the genera *Scaphiella*, *Opopaea*, and *Triaeris* that are brought to light mainly by sifting ground detritus.

In many ways the most interesting members of this series of hunting spiders are the tube weavers of the family *Segestriidae*. These cylindrical spiders, which retain the unpaired claws on all the tarsi, have the first three pairs of legs directed forward, and the front pairs, with which they hold their victims, armed below with numerous stout spines. A typical member is *Ariadna bicolor*, one-half an inch long, which is found almost everywhere in the United States; it has a purplish brown abdomen, and light brown cephalothorax and legs. Another genus, *Segestria* with several well-known, large species in Europe, is known from several species in California.

The retreat of *Ariadna* is a long, slender tube placed in a suitable crevice, with the silk continued outside and around the mouth opening as a silken collar. From the inner edge of the mouth originates a series of heavy lines that radiate outward like the spokes of a wheel, and that are attached at their ends a short distance beyond the collar. These radii, often two dozen or more, do not lie flat against the substratum, but are supported above the surface by little silken piers, one near the opening at the edge of the collar and the other out beyond the collar. The spider sits just within the tube, its six legs directed forward, in position to leap. The touching of one of the trap lines brings it out with surprising swiftness, like a jack-in-the-box, to the spot where the unlucky insect has tripped. *Ariadna* seizes the insect and instantly backs into the tube again. Even such formidable prey as a wasp is held almost helpless within the narrow tube—so narrow that the spider itself is unable to turn. A second, somewhat larger species, *Ariadna pilifera*, places its tubular retreats in the rock walls in southern Arizona and ranges from there south to the Isthmus of Tehuantepec in southern Mexico. A third species in our fauna is *Ariadna fidicina*, notable for its stridulatory device of coarse transverse grooves on each side of the head, which are activated by tubercles on the inner surface of the first femora. The sounds produced by this unique file and pick and the use to which they are put have not been identified so far.

In the remaining group of primitive hunters the chelicerae are soldered together at the base and along the inner edges and they cannot move separately. A single spiracle opening into the tracheal tube is placed far back toward the spinnerets, replacing the forward pair of openings of the previous series. The unpaired claw on the tarsi may or may not be present. Several different body forms and quite different habits are exhibited by the members of the five families, all but one of which are known from within our North American borders.

The primitive hunters of the family *Plectreuridae* are among the most generalized of all ecribellate spiders. They alone in this series still retain the full complement of eight eyes. They spin tubular retreats with small entrances fringed or ringed with silk, which are placed in dark situations of crevices in stone walls and bridges, in small ground openings, and under ground debris of all kinds. From these silken cells they move out at night to hunt. Only two similar genera are known. The genus *Plectreurys* has stout legs and the males have stout spurs on the tibiae of their front legs and probably use them for restraining or positioning the females during courtship and mating. Except for its smaller size, the male *Plectreurys* resembles the trap-door spiders in their stance and deliberate gait. The genus *Kibramoa* has thinner legs and the males have no coupling spur at the ends of the tibiae of their long front legs. The family *Plectreuridae* is exclusively North American with more than thirty species so far known from our arid southwestern states and adjacent Mexico, from scattered localities of eastern and southern Mexico, and from Costa Rica and Cuba.

The tube and net weavers of the family *Diguetidae* differ from the plectreurids in having only six eyes in a nearly straight row, and in having an expansive, spoon-shaped conductor in the male palpus. The single known genus, *Diguetia*, includes a small number of elongate spiders having bodies thickly covered with white hairs to form distinctive bands, and quite long legs ringed in black. The species of *Diguetia* suspend a long, vertical, tubular retreat, closed at the top and often three or more inches long, at the center of an expansive maze of threads. Within this web they move in the fashion of aerial spiders. The females incorporate their egg sacs in the tube retreat, laying one sac upon the other like the tiles of a roof. Over the long silken horn, which is widest and flared at the mouth opening, are placed leaves or plant debris available from the plant supporting the web or from the soil nearby. Cocoon retreats from different areas differ markedly in color, texture, and general composition. Favorite sites for the webs are the prickly pear and bush cacti of our arid regions where they may be found on almost any kind of shrub vegetation. Several species of *Diguetia* live in our southwestern states from Texas and Oklahoma to California and southward deep into Mexico. The genus is elsewhere now known from two species in Argentina, belatedly assigned to this family.

The sedentary weavers of the family *Loxoscelidae* are dull yellowish, six-eyed spiders, with long legs bearing only two tarsal claws. In many species the low carapace has the cephalic portion darkened to form a violin-shaped figure and this is responsible for one of their common names of "violin spider." The term "brown spider" more adequately describes the habitus of these spiders and is to be preferred. One of our species, *Loxosceles reclusa* (Plate 32), gained notoriety some years ago when it was shown to be capable of producing a cutaneous necrosis of considerable gravity to man. The bites of all species of *Loxosceles* are now believed to be venomous but only a few species have been investigated; their venomous status is mentioned in Chapter 11. The genus *Loxosceles* is represented in the United States by a number of common species; as many as fifty more are found in temperate and tropical South America and still others live in Africa and the Mediterranean region. The species of *Loxosceles* live in dark places where they spin a large irregular web with thick, quite sticky threads. Some favored stations are holes in the ground, caves, spaces under rocks and varied ground litter, and in buildings. Our *reclusa* is notably domestic, especially in the northern part of its range, and it is under such circumstances that most bites occur.

One family of this series, the *Sicariidae*, consists of quite large spiders that lie flat against large stones and hold their legs in the laterigrade fashion of the crab spiders. Instead of running away when disturbed, they rub the femora of their palpi against a file of grooves on the chelicerae to produce a sound much like the buzzing of a bee. These curious spiders occur chiefly in Chile and adjacent areas and in South Africa.

The species of *Scytodes* (Plate 32), our only representatives of the family *Scytodidae*, are mostly very pretty creatures with bodies less than one-half an inch long tinted in clear white or yellow, and delicately spotted and lined in black or more boldly marked with heavy dark spots or bands. The cephalothorax is oval and quite elevated, sometimes nearly globose; the abdomen is oval; the legs are very long and thin. The unpaired claw of the tarsi is usually present and of small size, but it may be completely absent. These nocturnal spiders live under stones, in rock fissures, under bridges, in houses, and even on the leaves of plants where they put down a thin flat web. The females carry the globular egg sac beneath the sternum, held in the chelicerae. Our several native species occur in our southern and southwestern states. The species of *Scytodes* have developed one of the most ingenious devices for capturing prey known among spiders, which the following example will serve to illustrate.

The spitting spider, *Scytodes thoracica*, handsome in a yellowish coat marked with small black spots, is a so-called cosmopolitan species that lives in some of our northern states. In habit it is domestic, and parades leisurely over the walls and ceilings of houses at night in search of small food animals. When one is discovered, *thoracica* gives a convulsive jerk of its body and squirts a viscous gum from its chelicerae, usually at a distance of one-quarter to one-half an inch. The victim is securely entangled and stuck to the surface by the gum, which is laid down by the rapidly oscillating chelicerae in ten, twenty, or more closely spaced, parallel bars. The spitting and entangling is almost instantaneous; thereafter the spider moves leisurely forward to claim its prey. The viscous liquid is produced in tremendously enlarged venom glands, which although given over largely to the production of viscous liquid, still produce a quantity of venom. When replete with viscous gum, the carapace of *Scytodes* is substantially elevated and globose but after spitting the carapace is substantially lowered.

11
Economic and Medical Importance

ECONOMIC IMPORTANCE

Spiders are among the dominant predators of any terrestrial community. When the fauna of the soil and its plant cover is analyzed, they come to light in vast numbers, in such convincing abundance that it is evident that they play a significant part in the life of every habitat. Working in the exceptionally rich jungle forest of Barro Colorado Island, Canal Zone, Eliot C. Williams estimated about 264,000 spiders per acre of the forest floor in a total fauna of 40,000,000 animals (nearly half of which consisted of ants and mites). While the fauna of the temperate zone has fewer *species* than the tropics, comparable habitats probably support a nearly equal numerical population. In 1907, W. L. McAtee found approximately 11,000 spiders per acre in woodland and 64,000 per acre in a meadow near Washington, D.C. From data given by Lucile Rice on animal fauna on the herbs and shrubs of woodland in Illinois in May 1934, I have estimated 14,000 spiders per acre, a number which would be considerably swelled by addition of the floor fauna. Yet substantial as these figures are, they are completely outdistanced by the total found in England by Bristowe in August 1938; he calculated that 2,265,000 were then present on a single acre in an undisturbed grassy area. Furthermore, Bristowe believes the average number of spiders per acre in all England and Wales is no less, and probably much more, than 50,000, and that the total spider fauna is not less than 2,200,000,000,000. Even when based on these conservative figures, the spider population of the United States would amount to an astronomical number.

The overall effect of such a large fauna of predators must be a very significant one. It should be remembered that spiders and insects have evolved together and that the latter have served largely as their food supply. Unfortunately, we still lack pertinent feeding data on spiders in most cases. By comparison with ants, certainly in the tropics, spiders are less important predators; but they are everywhere far more important than the highly considered birds in the number of invertebrates that they destroy annually. Spiders are ordinarily credited with catholic tastes and charged with eating all kinds of insects indiscriminately. This generalization is subject to many exceptions. Because such space-web spinners as the orb weavers and various sheet-web weavers concentrate on flying insects, it is probable that a higher percentage of their catch is made up of beneficial insects. On the other hand, some of the hunting spiders have been known to concentrate on obnoxious varieties. Much depends on the location of webs and the presence of wandering vagrants at a site where flights or emergences occur. Webs heavy with biting flies or annoying midges bear witness to the efficiency of spiders in helping to control economically destructive insects. One female black widow is reported to have destroyed 250 houseflies, 33 fruit flies, 2 crickets, and one spider during its lifetime. But in other locations similar webs may be filled with parasitic flies and others that are considered beneficial.

Although spiders are not usually thought of as being efficient agents of biological control, they have acted that role in a few instances. During 1923 and 1924 there was a tremendous increase in the numbers of bedbugs in Athens, particularly in the Greek refugee camps. Even when the inmates of the wooden barracks moved out into the roads they could not get rid of the pests, which followed their hosts with their usual persistence. Suddenly there came a rapid decrease and by 1925 the bedbugs had been eradicated. N. T. Lorando credited this success largely to the presence of the predaceous crab spider, *Thanatus flavidus.* He was much impressed by their efficiency in dispatching the bugs, destroying thirty or forty a day, and with its possible exploitation for systematic biological control. Later this spider was introduced into animal laboratory rooms in Germany by A. Hase, and again achieved great success in controlling bedbugs. Another member of this genus, *Thanatus vulgaris*, is often found in great numbers in warehouses in New York City, where it preys upon the many pests of stored cereals and other products. This spider of southern Europe and northern Africa has been introduced into coastal areas around our continent along with products from the holds of ships. It is especially abundant in the southwestern states and adjacent Mexico, where it was once thought to be endemic. In the main, however, too few experiments in the use of spiders for biological control have been made to indicate their possibilities in this field.

Economic entomologists have acknowledged the importance of specific spiders as control agents in certain cases. For example, in the Fiji Islands the cocoanut palm is ravaged by a moth that frequently occurs in tremendous numbers. During each outbreak, one of the large, strikingly ornamented jumping spiders, *Ascyltus pterygodes*, increases rapidly in number and attacks the caterpillars and pupae that survive the efforts of other controls. Again, various workers in America have identified spiders as important factors in checking cotton worms, gypsy moths, pea aphids, and many other destructive insects. They are especially effective in a prepared environment such as a cotton or cornfield. In any cultivated locale many kinds of insects will take up their abode, but the varieties that are detrimental to the plant crop are present in concentrated numbers. Spiders quickly overrun such areas and account for a considerable percentage of the larvae and adults of the pest. Both the vagrant species and the web spinners are important here. Their catch, when examined, is found to consist largely of the noxious insects.

If spiders are evaluated on the basis of their direct effect upon man in terms of nuisance, disservice, and usefulness, the conclusion is that they are essentially neutral. They litter our houses, but their unsightly, dust-catching cobwebs render us a distinct service in disposing of mosquitoes and flies that squeeze through our window screens. We find a use for their silk threads in certain types of optical instruments; the Papuan natives use the spider's matted webs for silken lures and fishnets. Although attempts have been made to rear spiders and take their silk for fabrics, the results have been unsuccessful. To the diet of some Laos and Siamese tribes the bodies of spiders add much needed fats and proteins, otherwise not obtainable. Spiders feed on many beneficial insects as well as on undesirable ones. On the other hand, their bodies provide food for game fish and for birds. And finally, the bites of a few spiders are poisonous. Thus one effect cancels out another.

MEDICAL IMPORTANCE: THE SPIDER'S BITE

Inasmuch as spiders are predators that normally specialize on insects and only rarely come in contact with human beings, their medical importance is not very great. Unlike their ubiquitous relatives, the mites, none of the spiders is parasitic on the bodies of man and his domestic animals. Furthermore, there is no evidence that spiders are the vectors of any of man's diseases. From time immemorial spiders have been used as charms to ward off disease, and they have contributed their bodies and silk for concoctions deemed of medicinal value. At the present time such primitive remedies are scorned, and we substitute instead, with like faithfulness, various patent medicines and an alphabet of vitamins.

Spiders once held an honored position among household remedies. The wearing of a spider in a nutshell hung around the neck was current in Longfellow's time, and brings to mind the more recent practice, at least in rural areas, of using asafetida or some other foul-smelling substance in the same way to ward off disease. One hundred years earlier, the belief was general that spiders, and their products, could alleviate many ailments. Indeed, this medical reputation produced a reasonable tolerance of house spiders. In rural communities it was believed that wherever they were abundant the human occupants enjoyed a relative immunity from certain diseases. Italian peasants still hold that cobwebs in stables are directly concerned with the healthiness of the cattle. Since spiders carry on such efficient warfare against stable flies, houseflies, mosquitoes, and other disease carriers, these old beliefs have some basis in fact.

Spider concoctions were administered by mouth or applied externally. One of the victims was little Miss Muffet who was dosed with spiders to cure many ailments. Warts and gout, constipation and jaundice, leprosy and all the communicable diseases, were treated in varying ways. Spiders were eaten alive or dead, were rolled up in pills of various kinds, were made into ointments to be rubbed on the body, and were brewed into liquors to be drunk. Less than a hundred years ago Mexican doctors prescribed as a specific a brew of alcohol and tarantulas.

The use of the silk of certain spiders to stop the flow of blood may still persist in rural areas of Europe and the Americas. References to the cobweb in this connection are frequent in literature, and indicate a wide application. The clean web of our domestic funnel-web spiders was placed over the wound much as sterile pads are applied and the numerous fine threads acted (largely in a mechanical way) to halt the flow. Unfortunately, ordinary cobweb is not sterile and its use sometimes resulted in infection.

Spider silk was also supposed to be of great benefit in the treatment of fevers. In 1821, N. M. Hentz commented as follows: "It has been found lately, that the web of a species of spider, common in the cellars of this country, possesses very narcotic powers, and it has been administered apparently with success in some cases of fever." Hentz named this spider *Tegenaria medicinalis* in recognition of this purported efficacy of its web. It is an abundant house spider of the eastern United States. In close proximity to it lives the cosmopolitan *Tegenaria domestica*, which is found in cellars all over the world and was once an important source of cobwebs for Europeans.

Deeply impressed in the minds of most people is the conviction that spiders of any kind are poisonous, and that many are deadly. It is this belief that keeps alive the distrust of these comparatively useful animals, a dread which in some cases produces hysteria at the mere sight of them. Many regard spiders on a par with poisonous reptiles and are always well supplied with tales of their size and virulence. Perhaps not so curiously, these same individuals will handle with small concern insects and animals far better equipped to do them injury. No spider is too small to be condemned as poisonous, and great size magnifies the reputation.

People imagine that they have been bitten by spiders when the actual culprits (*Honi soit qui mal y pense*) are fleas or bedbugs or biting flies. The responsibility for mysterious skin eruptions acquired during the night is often laid to a spider seen scurrying over the rug at the bedside, or serenely spinning its web in a corner of a room. It is not uncommon to attribute many types of dermatitis to spider bite. All mistakenly. What are the facts?

Spiders are shy animals that run away from pursuers whenever they can. Almost without exception, they will walk over the skin of man and make no effort to bite, regarding his body merely as a substrate. On the other hand, there are occasions when circumstances force them to attack. When held or squeezed they usually respond by attempting to bite. This is about all that can be expected of animals so poorly supplied with the higher sense organs that they cannot adequately see or comprehend just what man is. The most universally notorious spiders, the widow spiders, can be taken from the familiar security of their tangled webs and allowed to crawl over the hand, docilely unaware of the golden opportunity to use their venom.

Some few spiders have been charged with being vicious and even attacking man or animals without provocation. The Australian *Atrax* has such a reputation and is said to be quite belligerent. However, one wonders whether such spiders are equipped with eyesight sufficiently keen even to see a man at a distance of several feet. All spiders react to the presence of prey in an extremely efficient manner and, in running through the stereotyped actions of capturing and killing, they will give the impression of viciousness. Even their defensive attitudes, as witness that of the threatening but retreating tarantula with body and legs thrown back to expose glistening fangs, can be interpreted as indicative of belligerence.

By far the majority of spiders are relatively helpless creatures, always willing to scurry out of the way, never attempting to bite without the greatest provocation. Indeed, many of them must be forced by extreme means to bite when their venom is required for experiment. The larger spiders alone are capable of breaking the tough skin of a human being; the smaller ones can inflict mere superficial scratches which, in some cases, reach the small capillaries and draw a touch of blood. Ordinarily, there is no reaction beyond the slight laceration of the skin, and the sensation of pain is comparable to the jab of a pin. Only when the bite is inflicted by a large spider armed with strong weapons will there be any considerable injury to the skin. It must be remembered that a spider bite is always a pure accident.

The biting apparatus of the spider consists of the two chelicerae, and the venom sacs in which the poison is produced. Each chelicera has a stout basal segment, broadly articulated to the head, and a movable fang. When the spider bites, it presses the sharp, spinelike fangs into its victim and makes two separate punctures; at the same time, muscles squeeze the glands, forcing their poison into these wounds. The venom is usually a colorless liquid having the consistency of a light oil; it is said to have a bitter taste. The amount injected into the prey appears to be extremely variable—dependent on the available supply of the

moment, the age and condition of the spider, and the degree of excitation produced by the prey. There is reason to believe that the release of venom is to a
large extent controlled by the spider, and that in many instances the spider re-
frains from using poison on prey easily held in its grasp and not capable of
strong resistance. Repeated biting exhausts the venom supply; the bites become
progressively less poisonous.

The size of the spider does not give a clear index to the size of the chelicerae,
the volume of the venom glands, or the character of the venom. In two families
of distantly related spiders (*Uloboridae* and some *Liphistiidae*) the glands have
been nearly or completely lost; and in some others of closer kinship (*Scytodidae*
and *Filistatidae*) they are partially modified for other purposes and have become
tremendously enlarged. One of the largest American wolf spiders has glands
proportionately much smaller than those of the black widow and many lesser
spiders. The strongly built vagrants have a superior physical equipment, with
more powerful chelicerae and stouter legs to control their prey, and may get
along very well with a lesser amount of venom. More delicate spiders have the
problem of subduing large, often dangerous insects, and in some cases may compensate for their deficiencies by producing a greater or more potent amount of
venom. While there is no evidence to show that the quantity and virulence of
the poison are correlated with physique or other factors, it is clear that spider
venoms are variable and produce different effects.

Spider venoms are classified on the basis of their action on man and other
warm-blooded animals. Unfortunately, the various chemical compositions are
only now being made known. The toxins seem to be much more complex than
those of other arachnids, and produce symptoms showing the presence of elements affecting the blood, nerves, skin, and other tissues; several of these may be
present in the same venom.

The great majority of spiders, and almost all those from the United States and
other temperate areas, have a venom so feeble that its transitory effects are insignificant. In most instances, the bite is followed by local symptoms at the site
of the punctures—burning, throbbing, and similar painful sensations, numbness,
stiffness, and sometimes slight swelling. These symptoms usually persist for only
a matter of minutes, or a few hours at most, then disappear entirely—which indicates that the action is largely a local, mechanical one, and that the venom itself
lacks harmful toxins. The severity of this type of injury usually does not exceed
the sting of a wasp; only those individuals inordinately susceptible to the venoms
of arthropods are affected in any important way.

The poisons of a few spiders, however, are fortified with toxins that cause severe local or systemic reactions. Some contain toxins that destroy the cells in
the vicinity of the wound until large areas of cutaneous tissue are sloughed off,
exposing underlying muscles and organs. Such a progressive necrosis, often re-

sulting in gangrene, is caused by several South American wolf spiders, notably *Lycosa raptoria* of Brazil. Similar necrotic bites are caused by various species of brown spiders, the most dangerous being *Loxosceles laeta* of western South America, and several other species from the United States and other parts of the world. Other spider venoms have a special affinity for nerve tissue and inhibit the normal activities of important nerve centers by causing degeneration in the cells. These neurotoxins often strike quickly at the respiratory centers, causing severe systemic distress at points in the body remote from the site of the bite. The best known spiders with this type of venom are the widow spiders of the genus *Latrodectus*, but various unrelated ones from tropical regions are now credited with possessing similar poisons. It is of interest that the truly venomous spiders do not represent a single group, but include a few representatives from several distantly allied lines.

We now preface an analysis of the venoms of our American wolf spiders by mention of the large wolf spider of Europe that has taken its name from the city of Taranto in southern Italy. The reputation of this "tarantula" has persisted through hundreds of years, and around its venomous character and the peculiar methods identified with the cure of its bite has been built a vast literature of superstition and fiction. As McCook has written:

> When one is bitten by this spider, so the story goes, at first the pain is scarcely felt; but in a few hours after come on a violent sickness, difficulty of breathing, fainting, and sometimes trembling. Then he is seized with a sort of insanity. He weeps, he dances, he trembles, cries, skips about, breaks forth into grotesque and unnatural gestures, assumes the most extravagant postures, and if he be not duly assisted and relieved after a few days of torment, will sometimes expire. If he survives at the return of the season in which he was bitten, his madness returns.
>
> Some relief is found by divers antidotes, but the great specific is music. At the sound of music the victim begins the peculiar movements which are known as the "tarantula dance," and continues them while the music continues, or until he breaks into a profuse perspiration which forces out the venom. Thereupon he sinks into a natural sleep from which he awakes weakened, but recovered.

And then, quoting an older writer, McCook continues:

> Alexander Alexandrinus proceedeth farther, affirming that he beheld one wounded by this Spider, to dance and leape about incessantly, and the Musitians (finding themselves wearied) gave over playing: whereupon, the poore offended dancer, having utterly lost all his forces, fell downe on the ground, as if he had bene dead. The Musitians no sooner began to playe againe, but

hee returned to himselfe, and mounting up upon his feet, danced againe as lustily as formerly hee had done, and so continued dancing stil, till hee found the harme asswaged, and himselfe entirely recovered.

The spider credited with being the cause of tarantism is one of the large wolf spiders, *Lycosa tarentula*, which has been demonstrated by modern students to be no more virulent than comparable species of the genus. Various people have tested the notorious creature and reported that no ill effects result from its bite. The injury is similar to being jabbed with two needles. The pain is very sharp at first, but soon disappears, and the tiny wounds heal quickly without other symptoms. The reports of various Italian doctors have been very contradictory during the years of its notoriety, which is understandable when we realize that the real bites may have been caused by different spiders and that many purported ones were probably fictional.

There is no doubt that epidemics of tarantism swept southern Europe; they are matters of recorded history. However, the question as to what actually caused these demonstrations has not been fully answered, although there are several clues to their origin. As time went on, many doubters rose up to declare that the matter of the tarantula and tarantism was a fraud perpetrated upon gullible travelers who paid liberally to see the actions of supposed victims. Oliver Goldsmith declared that the peasants willingly offered to let themselves be bitten for the benefit of any tourist, and that the whole train of symptoms, and the style attending the tarantula dance were more or less in accord with the size of the fee paid by the onlookers.

It is quite probable that several things contributed to the outbreaks of tarantism. Some authors have suggested that it was a nervous disease, which attained epidemic proportions, then disappeared. An accidental and much less frequent variation could easily have been caused by the bite of the European widow spider "*malmignatte*," which has a neurotoxic venom capable of initiating serious symptoms. Indeed, this spider has been claimed by some workers to be the "tarantula." However, "*la malmignatte*" could not have been responsible for the great outbreaks of tarantism, which quickly spread over wide areas and claimed more and more victims through mass hysteria. A more interesting and convincing theory suggests that the dancing mania associated with the cure of tarantism is of religious origin. As T. H. Savory has written:

> Wherever the *tarantati* are to dance, a place is prepared for them, hung about with ribbons and bunches of grapes. The patients are dressed in white, with red green or yellow ribbons, those being favourite colours. On their shoulders they cast a white scarf, let their hair fall loose about their ears, and throw their heads as far back as possible. They are exact copies of the ancient priestesses of Bacchus. When the introduction of Christianity put a stop

to the public exhibition of heathen rites, the Bacchantes continued their profitable profession but were obliged to offer some irrelevant explanation. The local spider best supplied their needs.

Many large species of *Lycosa* occur in the United States, but not one has been singled out as being particularly venomous. They bite readily when handled carelessly, and in some instances the wound is as painful as a bee sting; but the effects disappear much more quickly. As noted, some individuals are abnormally susceptible to arthropod venom; upon these individuals the wolf spider may be able to inflict a wound of greater consequence.

Many of the spider bites in the warm lowland region around Sao Paulo in southern Brazil are ascribed to *Lycosa raptoria* and other large and abundant wolf spiders. At night these active vagrants frequently wander into houses and hide in clothing. While dressing in the morning, a person may be bitten on the hands or feet by the trapped spider, occasionally on the chest, abdomen, or other parts of the trunk. The toxic venom produces an extremely painful local lesion that sometimes spreads over large areas of the skin, reaching maximum severity where the skin is thick and the blood circulation relatively poor. Bites on spots well supplied with blood vessels—the blood seems to dissipate the effects of the venom quickly—rarely cause more than a passing injury. When allowed to run its course, the typical wound is difficult to cure, and will sometimes become gangrenous. Fortunately, a serum has been produced that alleviates the condition and brings on quick healing. In Peru, a similar type of necrosis is popularly supposed to be caused by the "pododora" (*Mastophora gasteracanthoides*), a fat, phlegmatic spider with two conical humps on its abdomen. The *pododora* lives in the vines of grape vineyards and is said to bite the workers when they gather the grapes. Inasmuch as the species of *Mastophora* are inscrutable introverts, exhibit little sign of life even when handled to excess, and have never been known to bite, the cause of the Peruvian necrosis must be some unknown spider or a different agent.

In 1934, Machiavello presented convincing evidence that the bite of *Loxosceles laeta*, the "*araña de los rincones*" or corner spider of Chile, was capable of causing a cutaneous necrosis of considerable gravity in man. The toxins in the venom destroy the cells in the vicinity of the wound, causing a black gangrenous spot (*mancha gangrenosa*) from which comes extensive sloughing of the skin and exposure of underlying tissue. In man the lesions vary in size from a small spot to large patches up to six inches or more in diameter. The eventual healing of the wound may sometimes require a hundred days. *Loxosceles laeta* moves about freely in the dark in homes and may hide in bed covering or clothing. Bites are most often suffered when the victims are putting on clothes in the morning and the wounds are centered mostly on the arms or legs. This dangerous spider, endemic to western South America, has been transported by commerce into many parts of that continent where it occupies mostly domestic situations in buildings. Of more recent date, infestations of *laeta* have become established in the buildings of Harvard University in Cambridge, Massachusetts, and in three cities of Los Angeles County, California. Efforts to eradicate these infestations have not been successful. Thorough housecleaning with special attention to dark corners, closets, attics, and the like are recommended procedures. Up to the present time no bites of *laeta* have been reported in the United States and efforts to control the infestations continue.

It seems clear that all species of *Loxosceles* are venomous even though up to now only half a dozen have been so identified. In 1957, Dr. Curtis W. Wingo and associates at the University of Missouri demonstrated that *Loxosceles reclusa* (Plate 32), widespread in our midwestern states from Illinois through southern Texas, produced the same kind of necrosis as that of *laeta*. This spider quickly gained a wide notoriety in the media, became the subject of innumerable investigations of its range, ecology, and especially its venom, and acquired various popular names of which "brown spider" and "violin spider" were most appropriate. Our brown spider *reclusa*, found outdoors in natural situations as well as in buildings, becomes increasingly limited to houses as one travels North. Single specimens of *reclusa* have been transported by trade and travel to such distant places as California, Arizona, Minnesota, New Jersey, and Florida, well outside its natural range.

More than a hundred bite cases for *reclusa*—these also usually occuring in houses while the victim is putting on clothing—have been reported from the United States. The progression of symptoms from the early black spot to a depressed ulcer takes much the same course as that of *laeta*. The venom of *reclusa* is far less potent than that of *laeta*. About fifty deaths have been charged to *laeta* in South America (some of these due to secondary toxins in the venom causing blood poisoning and kidney failure) as compared with less than half a dozen deaths in the United States. Some bites in our far western states are attributed to *Loxosceles deserta* (formerly *unicolor*) and *Loxosceles arizonica*, as reported by Dr. Findlay E. Russell and his associates. In the West, brown spiders are often abundant in natural habitats under ground objects, in litter around buildings, and occasionally in houses. Only fifteen certain or likely cases of envenomation by *Loxosceles* in southern California were recognized by Dr. Russell and his associates during a fourteen year period.

The number of spider bites brought to the attention of doctors each year in the United States must be very large; in southern California alone some 400 bites are reported during each year. Many other bites produce insignificant symptoms and are never reported. Others cover a wide range of severity and may culminate in hospitalization with dangerous symptoms. Only in a small percentage of cases is the offending spider or arthropod apprehended as evidence. Great care must be exercised by the doctor in evaluation of the symptoms without presence of the biter. As pointed out by Dr. Russell, a whole series of arthropods, including fleas, ticks, bedbugs, ants, biting flies, and kissing bugs, may produce lesions that resemble those of *Loxosceles* envenomation. Spider venom poisoning by multiple biters collectively is called *araneidism* or *arachnidism* by American workers. About eight people each year, mostly young children bitten by widow spiders, die from this disease.

The *Loxosceles* characterized above and our several species of *Latrodectus* are our truly dangerous spiders already with significant medical histories. In addition, about fifty different species of spiders have been implicated in bites on humans in the United States; and this number could be greatly enlarged if all the biters were apprehended and classified. Among these lesser offenders are one or several species of the following genera: *Phidippus, Thiodina, Drassodes, Liocranoides, Chiricanthium, Peucetia, Lycosa, Neoscona, Argiope, Araneus, Steatoda, Ummidia, Bothriocyrtum,* and *Aphonopelma.* The first seven are vagrants and running spiders, the next three are orb weavers or cobweb spiders, and the last three are mygalomorph spiders. The symptoms from bites of most of these quite different biters are slight to acute initial pain, some redness and local swelling, and transitory illness usually without necrosis or systemic reactions. Only occasionally do symptoms persist or require hospitalization. It must be remembered that each victim responds to the venom in terms of his unique physical and allergic makeup, and this makes for wide differences in the clinical picture. The genus *Chiricanthium* perhaps can be sorted out of the above series as one with a venom of medium intensity. Three species from the United States have been mentioned as biters: our widespread *inclusum*, living mostly on vegetation outside but also found in houses; *mildei*, introduced into the New England area about thirty years ago and now becoming widely distributed as a house spider; and *mordax*, brought from Australia into the Hawaiian Islands and now common in buildings. The venoms of these spiders have been described as causing acute pain followed by mild to intense systemic reactions and then a return to normalcy within a single day. A mild necrosis may persist for a few days. The European *Chiricanthium punctorium* has been characterized as, next to *Latrodectus tredecimguttatus*, a venomous spider but of medium importance.

The above generalizations apply only to the spiders of minor toxicity in the United States and perhaps also of Europe. The same genera and many different ones occurring in tropical and distant areas may have distinctive and dissimilar venoms. Again it can be stated that our dangerous spiders come from widely separated groups. The most venomous species of Europe and North America are widow spiders. The Australian widow *Latrodectus hasselti* has been responsible for a few deaths but their spider with the darkest reputation is the funnel-web spider *Atrax robustus*, a belligerent biter (most often a male) with powerful neurotoxic venom responsible for many deaths of children within a few hours. The most dangerous species of eastern South America are the aggressive, wandering ctenids of the genus *Phoneutria*, principally the species *fera* and *rufibarbis*, which abound in the Sao Paulo area of Brazil. More than 200 victims are treated each year in the hospital of the Instituto Butantán. The large spiders come into houses and inflict painful bites often fatal to children under seven years of age.

The great spiders that Americans know collectively as tarantulas are capable of inflicting a deep, painful wound with their chelicerae. Because of their size and hairiness, they are feared, and the usual reports credit them with being extremely dangerous. Dr. William J. Baerg of the University of Arkansas has studied this group of spiders for many years. He has concluded that no species from the United States is able to produce anything more than trivial symptoms in man, indeed little more than the mechanical injury of breaking the skin. Considerable pain often accompanies the bite, it is true, but this ordinarily lasts less than an hour. Our tarantulas live such secretive lives that opportunity to bite man does not often present itself. Only in the summer and fall of the year are they to be seen in numbers, and those seen are mostly males wandering about in search of females in their burrows. The males are not very belligerent, prefer to run away when disturbed, and are rather easy to tame.

The effects of the bites of common North American tarantulas on laboratory animals vary considerably. In some instances the bites seem to have no noticeable effect on white rats and guinea pigs. In other cases these small mammals die quickly. The large tarantula of the Panama Canal Zone, *Sericopelma rubronitens*, kills guinea pigs in half an hour, and causes painful symptoms in man that persist for several hours. Some *Dugesiella* and *Aphonopelma* species of similar size produce no adverse symptoms on experimental animals or on man. *Ummidia* (formerly *Pachylomerus*) *audouini*, our largest eastern trap-door spider, was allowed to bite man. Although this spider is as large as many of the tarantulas, its venom was found to be of less potency than that of many common true spiders. Indeed, there is little reason to believe that any of our mygalomorph spiders are dangerous to man.

However, since the large tarantula group is composed of quite diverse elements from many parts of the world, we cannot label them all harmless without much more data. The poison of the largest spiders in the world, immense creatures of the genera *Lasiodora* and *Grammostola* from Brazil, is highly toxic to cold-blooded animals, but is nearly ineffective on warm-blooded animals and man. The mere mechanical hurt from the fangs of such huge spiders is considerable; and this, along with local complications not definitely due to the venom, has given them a reputation that they do not entirely merit. On the other hand, the venom of *Trechona venosa*, a large funnel-web tarantula from South America, vastly inferior in size to many true tarantulas, belongs to the neurotoxic type and was found to be dangerous when injected into human beings.

In the United States the only spiders known to have a dangerous venom producing neurotoxic symptoms are the "widow" spiders of the genus *Latrodectus*, a name derived from the Greek meaning a "robber-biter." The widows occur around the world in the tropics, and extend far into the northern and southern temperate zones. The genus comprises twenty or thirty mostly jet-black spiders remarkable for their beautiful red markings and all notorious for their venomous properties. Because of their great variability in color patterns, they have received a plethora of scientific names; and they have been singled out, given expressive common names, and indicted by peoples from widely separated regions of the world.

In southern Europe and Africa bordering the Mediterranean lives *Latrodectus tredecimguttatus*, the *malmignatte* of Corsica and Italy, gaily splashed with red and generally feared by the peasants. Farther to the east, this same species is mostly black; it becomes the "karakurt" or "black wolf" of Russia. Other distinctive species are known by various names in Arabia, the Gulf of Persia, and in northern Africa. Much farther to the east in India and Malaya is found *Latrodectus hasselti*, whose dorsum is marked by a broad crimson stripe running down to the tip of the abdomen. Under various names, *hasselti* is found throughout the major islands of the East Indies, and extends down into New Guinea and Australia, where it has been dubbed "red-back spider." On the beaches of New Zealand lives *Latrodectus katipo*, a dusky congener known as *katipo*, the night-stinger of the Maoris.

In southern Africa is found another species of the genus, *Latrodectus indistinctus*, which has the jet-black abdomen marked above with small white spots, and which is known as the *knoppiespinnekop* or "shoe-button spider." Another black species of Africa is the *vancoho* or *menovadi* of Madagascar, *Latrodectus menovadi*.

In North America, a principal species is *Latrodectus mactans*, a name misused to cover several distinctive species of the world but now limited to a species of eastern North America, eastern Mexico, and the West Indies. Our *mactans* became the most feared and notorious of all the widow spiders. In the United States it is called "black widow" (an inappropriate name but one firmly established, having reference to the erroneous belief that the female invariably kills the male following the mating), and also the "hourglass" or "shoe-button spider"—common names describing the shape of the red ventral marking and the form of the jet-black abdomen. In the West Indies it is called the *cul rouge* or *veinte-cuatro horas*. In South America our *mactans* is replaced by a series of still little known species given local names; in Peru it is *lucacha;* in Chile *guina* or *pallu;* in Bolivia *mico;* and in Argentina *araña del lino*. In Mexico *viuda negra* is largely replacing the more interesting *araña capulina* of the Mexicans and the *chintatlahua* of the Indians.

At least five well-marked species of *Latrodectus* occur within the United States. All are quite large, subglobose theridiid spiders (very large females sometimes measuring a little beyond one-half an inch in length) in which the lateral eyes of each side are widely separated, a condition nearly exclusive to this genus in its family. In addition to *mactans*, our species *variolus* and *hesperus* deserve to be called "black widows" inasmuch as the adults of all three are mostly jet-black and more or less marked with crimson spots. When the spiderlings leave the egg sac they are whitish and progress through a gradual series of stages in which black gradually covers most of the white, yellow, and red spots and bands. In adult females the dorsal markings on the abdomen are frequently lost, leaving it almost all black; on the venter the red hourglass marking is rarely lost. The males retain much of the bright color pattern of the immature stages and are much smaller than the females.

Our three black widows are much alike, present much variation in size and coloration, and are often difficult to separate. Each can be characterized briefly as follows: The southern black widow, *Latrodectus mactans* (Plate 9), ranges north to southern New England; the cephalothorax and legs are usually all black; the abdomen is shiny black with a red spot just above the spinnerets and sometimes red spots or a band extending forward; the red hourglass marking on the venter is usually present and entire; the quite small males are black with red and white spotting on the abdomen; this black widow frequently lives in disturbed areas where its irregular web is placed under ground detritus near buildings. The northern black widow, *Latrodectus variolus* (Plate 10), ranges widely in the eastern United States south into western Florida and eastern Texas; females are much like *mactans* but typically have a central row of red spots on the abdomen; the ventral hourglass is divided into two transverse bands; males are somewhat larger and often have the legs marked with paler rings; in Florida this black widow lives in moist areas and places its webs in the branches of trees, but its habitats in other parts of its range are more varied. The western black widow, *Latrodectus hesperus* (Plate 10; Plates XI–XIV), ranges widely from Oklahoma, Kansas, and middle Texas throughout the southwestern states; females have the cephalothorax and legs black; the abdomen is typically all black above, more rarely with some paler spots of the immature pattern present; the ventral hourglass is most often complete, somewhat variable, and occasionally divided; the males are typically much paler than those of the other species, with bands and spots on the cephalothorax and abdomen subdued to give a general gray appearance; this black widow occupies a wide range of habitats over several life zones, most often placing its webs near the ground under objects, in animal burrows, under bridges, frequently on shrubs above the ground or even in trees.

The red widow, *Latrodectus bishopi* (Plate 11), is one of the most colorful members of our spider fauna and is found only in sand-pine scrub associations in central and southeastern Florida. The carapace and legs are bright orange or reddish and the abdomen, which rarely is all black, is usually brilliantly spotted with red or yellow marks on the dorsum, and these are retained by the adult female. The pale ventral marking is a transverse band or triangular spot, representing at most half of the normal hourglass markings of many other species. The red widow places its web three or four feet above the ground, spreading the lines from frond to frond of the palmettos to form a snare reminiscent of the sheet-web weavers. "The web retreat is made by taking a palmetto frond and rolling it into a cone. The interior of the cone is lined with silk and the eggs are hung from the sides of the cone. The egg sacs are light gray to white in color and have a fairly soft outer covering unlike the papery covering of *L. mactans* and *L. variolus*." There have been no published accounts of the poisonous nature of this beautiful spider, but there is little doubt that its venom will be found different from those of its close relatives.

The brown widow, *Latrodectus geometricus* (Plate 11), is widely distributed around the world in the tropical zone, and is the dominant species over most of Brazil and in the eastern coastal portions of South America. Although it now ranges widely, having been brought in by commerce and established in many parts of the world, its occurrence in North America is sporadic. In the United States it is established only in southern Florida where it lives mainly in domestic situations in or near buildings in the fashion of the house spider *tepidariorum*. The brown widow is ordinarily grayish or light brown, mottled with blackish, red, and yellow markings, but on occasion it is black, thus assuming the more typical coloration of the group. The hourglass on the venter is dull orange or reddish in color. The unkempt webs resemble those of *mactans* and in them are stationed spherical egg sacs covered with scattered tusks of silk. Although the venom is potent and there are some records of bites on man, the brown widow seems to be the least dangerous of our species.

The typical black widows are tangled web weavers that spin a rather small snare of coarse silk in dark locations. The core of the web is a silken tunnel in which the spider often spends the daylight hours, and into which it retreats when disturbed. Radiating from the tube are numerous strands forming an irregular mesh. The whole web may be limited to the space in a small burrow, but an aerial snare usually projects in all directions for a few inches to a few feet. The silk is heavy and strong and can entangle the largest terrestrial insects that blunder into it. In most instances the webs are placed in or close to the ground. The abandoned burrow of a rodent may be appropriated and fitted to the needs of the spider. A recess under a stone, a crevice in a dirt bank, the spaces under chips of wood, log piles, stone heaps, or stacked material of any kind, afford excellent sites for the webs of *Latrodectus*. Indeed, man provides so many excellent stations for webs that these unwelcome spiders abound near his habitations. In the eastern United States they are almost invariably associated with littered areas, with the dumps of large cities, and with garages, barns, storage sheds, and privies. Indoors, in addition to occupying dark crevices, their webs are placed in the angles of doors and windows and behind shutters. Although reputed to live inside houses and they often do, they are not found there as frequently as is generally believed. Our western black widow uses a wider range of habitats than do our eastern species. They are reported to live in birds' nests in pine trees, and to infest grape arbors in Colorado. In arid parts of Arizona almost every crevice in the soil harbors a black widow, and their nests are often found in cholla cacti and agave plants. Every state in the union and some of the Canadian provinces have endemic black widows.

Our black widows are shy sedentary creatures, largely nocturnal in habit, which live retiring lives in the small world of their coarse web. The females rarely leave the snare voluntarily and are clumsy creatures when not in intimate touch with the lines. Much more venturesome are the males which, only about a third as large as their mates, must search for them during the brief mating season. In this weaker sex the chelicerae are very small, and the males are reputed never to bite. To the female credit must be given for all poisonous injury to animals and man.

In spite of great reputation, the females are timid. They ordinarily make no effort to bite, even when subjected to all kinds of provocation. They are never aggressive and make no effort to attack, preferring instead to retreat or lie perfectly still. The danger lies in the fact that, because they live in abundance near man, they may be accidentally squeezed against his body in some way. They lie hidden in the folds of clothing, in shoes, or under objects in corners. When a body contact is established under normal or exceptional circumstances, the black widow bites in self-defense. It is in the old-fashioned outdoor privies that the black widow has been particularly dangerous. Ideal sites for nests and webs are found under the seats, and a large fly population provides plenty of food for the spider. The threads of the web often fill large areas beneath the seats and they are sometimes touched by parts of the body when the houses are in use. A gentle brushing of the web initiates the normal response of the spider to the presence of insect prey; it rushes to the site and bites the object vigorously, treating it in the same way as it would a large insect. Most victims of this type of black widow poisoning are males, and the injury is centered on the external genitalia.

The venoms of *Latrodectus* species from all over the world are highly potent and bring on similar symptoms in animals and man. Much of the pioneer work in this field was done by Dr. Emil Bogen of California, in a state where a high percentage of the bites occurred. We now know that the black widow he dealt with was the western *hesperus*, not the eastern *mactans*, but the details of the clinical picture are largely the same. The bite of *hesperus* causes an initial trivial pain comparable to the prick of a needle, and leaves two tiny red marks at the site of entrance of the two fangs. Almost immediately there develop sharp local pains which quickly reach maximum intensity mostly within half an hour; these may lessen or be persistent for a number of hours. Indeed, sharp pain is a prime symptom of the bite; it has been described as "intense, violent, agonizing, exquisite, excruciating, griping, cramping, shooting, lancinating, aching and numbing, and was either continuous and incessant, or paroxysmal and intermittent." The pain moves gradually from the wound area to other parts of the body, and finally concentrates in the abdomen and the legs.

In addition to the intense pain occasioned by the neurotoxic venom, many other symptoms have been described, most of them being consequences of the direct action of the venom on the nerve centers. Nausea and vomiting, faintness, dizziness, tremors, loss of muscle tone, and shock are all systemic symptoms frequently noted. There may be speech disturbances and general motor paralyses of various kinds. When the respiratory centers are strongly affected, there follows difficulty of breathing, cyanosis, and prostration.

Many different remedies have been used in the treatment of black widow poisoning. Most of them have not at all changed the course of the symptoms, and some have undoubtedly made the condition more serious. Alcohol is now known to have a most harmful effect, and its use at any time during the course of the disease is a serious mistake. It is recommended that all patients under sixteen years of age or over sixty, or with hypertensive heart disease, or with symptoms and signs of severe envenomation, be hospitalized as quickly as possible following the bite. On admission one ampule of antivenin (*Latrodectus mactans*) (Lyovac) should be given intravenously in 10 to 50 ml of saline solution after appropriate skin or eye tests. Young children should be watched carefully during the first ten hours of hospitalization. The intense pain will be alleviated by the drugs favored by the attendant doctor. Hypertension and muscle spasticity can be overcome by the intravenous injection of calcium gluconate, still an effective remedy for arresting symptoms since World War II.

Whereas the illness following black widow poisoning is frequently grave, a fatal outcome is rare. In minor cases the symptoms, even in untreated patients, rarely persist for more than two days. Very young children and older people are more likely to be seriously affected than those in more robust condition. In 1794 Luigi Totti in Italy described the death of a five-year old child in less than twenty-four hours following the bite of the *malmignatte*. In older people death is often due to complications rather than to the venom itself. The tiny wound may allow entrance to germs causing tetanus, erisipelas, or other dangerous diseases. The additional strain on the circulatory system occasioned by the venom may cause cerebral hemorrhage. These secondary causes of death, however, do not minimize the importance of the spider venom in initiating the condition leading to the fatal consequence.

Approximately 1,300 cases of black widow bite (chargeable to our three different species) were reported in the United States from 1716 to 1943. Every state was represented on the list, but nearly half the total, 578, were from California. Virginia led the eastern states with 173 cases; Florida had 126. Single instances from many far northern states, such as Maine, Vermont, and Minnesota, reflect the paucity of black widow fauna in those areas. A total of 55 deaths were recorded, about 4 percent of the total number of bites. This percentage of fatality is low, but would have been even lower had all the bite cases been available for the record. Many people who are bitten are not sufficiently affected to receive medical treatment; therefore, while fatal cases are usually reported, nonfatal ones do not often find their way into the record. It is also known that some of the deaths were the result of improper treatment, such as the administration of alcoholic potions, or even abdominal operations performed by physicians who erred in their diagnosis.

Without wishing to belittle the importance of even a single fatality from these dangerous spiders, it is nevertheless obvious that their medical significance has been overemphasized. In 1934 these spiders became notorious overnight when an intemperate press sank to ludicrous depths in disseminating its absurd exploits to a gullible public. The wave of notoriety which made this spider Public Enemy Number One has now subsided, and the hourglass spider has retired to an obscurity in keeping with its slight importance in the lives of most North Americans.

The high toxicity of the black widow's venom is now well established. The claim that it is among the most highly toxic of all venomous creatures is probably correct. As compared with the venom of the prairie rattlesnake, which is largely hematoxic in character, it is about fifteen times as potent on a dry-weight basis. Comparison on the basis of such animal tests must end there since it is clear that the rattlesnake is much more dangerous than the black widow. Each year about 1,500 snakebites from all venomous varieties are recorded in the United States, and approximately 5 percent result in fatality. Thus, in a single year poisonous snakes account for as many bites and kill more people than are credited to the black widow in more than 200 years. The average person's chances of dying from snakebite are about the same as being struck by lightning. The chance of dying from a spider bite is considerably less.

Looked upon objectively, it becomes clear that the importance of the black widows and brown spiders is not sufficiently great to warrant designation as a menace in any part of the United States. That does not mean that we should look without concern on abundant and dangerous aggregations of these poison-ous spiders in our environment. The domestic habits of these biters make control precautions desirable. It is generally believed that systematic and frequent eradi-cation of the spiders and their egg masses by mechanical means is the most satis-factory method of control. At night the spiders can be easily detected with the aid of a headlamp or flashlight and destroyed by hand. A reasonable amount of neatness in the storage of equipment and frequent disposal of rubbish will re-duce the available sites for nests and webs inside and outside the house. A peri-odic examination of outdoor privies (many still exist) should be made, and the undersides of seats should be painted with creosote, crude oil, or some other re-pellant. Various insecticides are available to supplement mechanical means of control. These precautions are especially important for people in the southern and southwestern parts of our country where large populations of these spiders are often found. Finally, everyone should be able to recognize our black widows and brown spiders by sight and avoid contact with them.

12

The North American Spider Fauna

The area covered in this book comprises much of the North Temperate Zone of the New World, the Nearctic realm. This is one of the natural biological land areas of the world and includes the part of North America north of the tropical region of Mexico and Central America. It is a vast land mass characterized over much of its surface by a climate that may, with certain reservations, be termed temperate. The present southern limit is a tropical climate that bars as effectively as an ocean the southward extension of the northern faunas. To the north, the faunas gradually become diminished as they approach the pole, being greatly reduced where conditions of extreme cold persist for long periods, and almost completely lacking on areas of permanent glacial ice.

The North American region has maintained its general form and its separation from other major land areas of the world for vast periods of time, probably since the Paleozoic era. Its isolation has been accomplished to the east and west by broad oceans, to the south by the tropics and transitory ocean gaps, and to the north by arctic wastes. Whereas physical isolation has for the most part been complete, there have been periodic joinings to adjacent land masses by means of broad bands in the south and narrow bands in the northern reaches. Animals have moved northward into North America from centers in South America, and vice versa. Interchange of faunas has been effected between Alaska and Siberia by the Bering Strait land connection, a bridge believed always to have been a relatively narrow one that allowed animals to pass in both directions when climatic conditions were favorable.

The result of this faunal intercourse during the Tertiary period is reflected in the similarity of the faunas of the temperate zones of both the Old and the New Worlds, which together constitute the Holarctic realm. Before the ice ages, the polar region probably enjoyed a much milder climate, which made possible an intermingling of faunas of the whole northern belt around the world. Over a large part of North America the climate was subtropical, and many tropical forms penetrated far into the north. During Florissant time in the Oligocene, Colorado had a climate at least as mild as that of our southeastern Gulf States, and had as part of its fauna a silk spider (*Nephila*) perhaps identical with the one now living in Florida and some of our southern states. The spiders of America and Europe were probably quite similar at that time, a conclusion reached through study of a very imperfect fossil record and perhaps not entirely valid. The similarity between these faunas, however, is not a recent one. During the Paleozoic era, the same types of primitive spiders lived in Europe as those found in the Carboniferous slates of Illinois. It is often postulated and perhaps correctly that the much richer spider fauna of Paleozoic Europe points to the derivation of American forms from that or some other Eurasian center of origin. In the splendid amber fauna of the Oligocene era in Europe there is quite likely a picture of the wealth and variety of our own American spider fauna for the period, even though a close relationship on the basis of existing fossils cannot be demonstrated. An interchange of new types between the Old and the New World has gone on almost continuously for more than 400 millions of years, interrupted for brief periods by transitory barriers which were not too great for crossing by tolerant and enterprising spiders.

The present faunal kinship between the temperate zones of the Old and the New World is a striking one, which reflects itself in the identity or close relationship of most of the genera and many of the species. Unfortunately, the faunas of the two regions are not well enough known to make possible explicit comparisons. The number of known species of spiders from the entire Holarctic realm is around 6,000 or 7,000, a number far below the real total—possibly no more than half of it. In Palearctica, only the European spiders have been well studied; vast expanses of temperate Asia are almost unknown. However, it is believed that the fauna is a very homogeneous one, and is modified longitudinally only by the character of the tropical genera and species that press northward into the temperate zone for varying distances. In Nearctica, only the spiders of the northeastern United States are well known; knowledge decreases progressively as one leaves that area.

One finds the same types of spiders in England and northern Europs as occur in Siberia and Japan, and in Canada and the United States. None of these regions shows a marked superiority in the number of species from comparable ecological zones; rather, they are essentially equal in faunal wealth, indicating that similar biological areas usually support quite similar kinds and numbers of animals. The following totals are only approximations of the present known numbers: In Palearctica there are about 3,500 species, of which 557 occur in the British Isles, 450 in Spain, 688 in Switzerland, 341 in Norway and Sweden, 391 in Greece,

1,335 in France, and 3,100 in all Europe; in Nearctica, there are about 3,000 species, of which 50 occur in Greenland, 249 in Alaska, 497 in Georgia, 600 in New York State, 650 in New England, and more than 900 in California. France has long been a center of arachnology, so it is not surprising that its spider fauna is so well known. This is also true of the British Isles, from which the number of known species is near the real total for the area. Intensive studies of American and Asiatic spiders will ultimately bring the faunas of comparable regions to parity with Europe.

One index of relationship is the number of American spiders that are the same as those from the Palearctic region. The list given below has now reached a total of 217, and this number can be expected to increase with more collecting and critical studies. A high percentage of this total represents a panboreal or temperate Holarctic fauna of long-established residents that have lived in this zone around the world for thousands of years. Few of them have changed sufficiently to be called subspecies. Through the medium of the ballooning threads, many strictly boreal spiders have been able to keep in quite intimate contact with their own kin in Alaska and Siberia. There is little reason to doubt that aerial spiders, restricted only by prevailing winds, can cover great distances at high altitudes, and live through the ordeal in sufficient numbers to establish themselves. The fact that relatively few kinds may become established demonstrates that mere access to a new region is not enough. The immigrant must be able to cope with a complex climate possessing numerous characteristics, any single factor of which may be capable of excluding it from survival. Most boreal spiders are anciently American and have not come in through accidental introduction of man. *Araneus marmoreus* and *nordmanni* are quite as typical of America as of Europe, and the American derivation of these is quite as plausible as a European one.

The ubiquitous spiders have been transported and distributed largely by man in his ships. They do not represent any special group, but single species from a number of different genera derived from widespread sources have become specialized to live and prosper in many parts of the world without regard to differences of climate. One of the most successful is *Achaearanea tepidariorum*, a house spider which is especially common and widespread. The remaining so-called cosmopolitan species (identified by the letter "C" on the list) are far less widely distributed, and are often rare or entirely lacking even in apparently suitable regions. Some other spiders have become widely disseminated into several continents in tropical belts around the world. Some of these tropicopolitan species (identified by the letter "T" on the list) barely impinge on southern portions of the Holarctic realm.

Transportation of plants and animals by man in ballast and goods of ships, by airplanes and land vehicles of all kinds has been going on constantly since settlement of North America. Spider immigrants came in very early, at first chiefly from Scandinavia and the British Isles, and later from a plethora of sources. Although this introduction was probably widespread, there materialized two principal centers where friendly climates and acceptable animals came together. These were the northeastern coast from Newfoundland through New England and the Pacific coastal region of British Columbia and Washington. Recently the delta region of Louisiana and Mississippi have become important for a different fauna of introductions. Such domestic species as *Achaearanea tepidariorum*, *Pholcus phalangioides*, and *Tegenaria domestica* quickly established themselves in a new climate and expanded their ranges widely into available habitats. Our *Zygiella x-notata* and *atrica*, first noticed about 1880 on wharves in Massachusetts and likely brought in from northern Europe on shipments of trees, by 1911 had become abundant on coastal Massachusetts and Rhode Island. Many immigrants are known to have arrived, or better said to have been noticed only recently. *Pholcus opilionoides* was found in cellars and buildings about forty years ago and continues to occupy such habitats in the New York region. Even more recent records are those for the European *Ozyptila praticola* and *Clubiona lutescens* in the Washington–British Columbia enclave. *Steatoda bipunctata*, probably a very old immigrant, is gradually moving westward from Nova Scotia and New England and finds buildings and adjacent trash heaps favorite retreats. Such habitats near Toronto, Ontario, used to be occupied by our American *Steatoda borealis*, but these are now usurped by the foreign *bipunctata*. The situation is much the same for *Chiricanthium mildei* which in about thirty years became the dominant house spider of its class in the New York and Boston areas, and essentially excluded our native *inclusum* from what was one of its alternate habitats. Many European spiders have arrived in this country and colonized successfully, and some of these have reduced or crowded out native species from special habitats. On the other hand, very few American spiders have moved successfully into the European fauna, a generalization that applies quite as graphically to plants and animals of all kinds. Single specimens of *Eperigone maculata*, found in a hothouse in Switzerland, and *Coriarachne versicolor* from southwestern France, have been reported. Our most notable invaders are *Psilochorus simoni*, now widespread in wine cellars and buildings from England to Italy, and *Eidmannella pallida*, quite recently introduced into England where it favors hothouses and recesses under stones.

Uloboridae

T *Uloborus geniculatus* (Olivier)

Oecobiidae

C *Oecobius annulipes* Lucas
Oecobius cellariorum (Duges)
Oecobius putus O. P.-Cambridge

Dictynidae

Dictyna annulipes Blackwall
Dictyna arundinacea (Linnaeus)
Dictyna major Menge
Tricholathys lapponica (Holm)

Amaurobiidae

Amaurobius ferox (Walckenaer)
Amaurobius similis (Blackwall)
Arctobius agelenoides (Emerton)
Ixeuticus martius (Simon)
Metaltella simoni (Keyserling)
Titanoeca silvicola Chamberlin and
Ivie

Loxoscelidae

C *Loxosceles rufescens* (Dufour)
C *Loxosceles laeta* (Nicolet)

Scytodidae

T *Scytodes fusca* Walckenaer
T *Scytodes longipes* Lucas
C *Scytodes thoracica* (Latreille)

Dysderidae

C *Dysdera crocata* C. L. Koch

Pholcidae

Holocnemus pluchei (Scopoli)
Pholcus opilionoides (Schrank)
C *Pholcus phalangioides* (Fuesslin)

T *Physocyclus globosus* (Taczanowski)
Psilochorus simoni (Berland)
T *Smeringopus elongatus* (Vimson)
C *Spermophora senoculata* (Duges)
(syn. *meridionalis* Hentz)

Mimetidae

Ero furcata Villers

Nesticidae

Nesticus cellulanus (Clerck)
Eidmannella pallida (Emerton)

Theridiidae

C *Achaearanea tepidariorum* (C. L.
Koch)
Crustulina sticta (O. P.-Cambridge)
Ctenium lividum (Blackwall)
Dipoena hamata Tullgren
Enoplognatha ovata (Clerck)
Enoplognatha tecta (Keyserling)
Enoplognatha thoracica (Hahn)
Latrodectus geometricus C. L. Koch
Neottiura bimaculata (Linnaeus)
Steatoda albomaculata (De Geer)
Steatoda bipunctata (Linnaeus)
Steatoda castanea (Clerck)
Steatoda erigoniformis (O. P.-
Cambridge)
Steatoda grossa (C. L. Koch)
Steatoda triangulosa (Walckenaer)
Theridion berkeleyi Emerton
Theridion impressum L. Koch
Theridion melanurum Hahn
Theridion ohlerti Thorell
Theridion ornatum Hahn
Theridion petraeum L. Koch
T *Theridion rufipes* Lucas
Theridion simile C. L. Koch
Theridion tinctum (Walckenaer)

Theridion varians Hahn
T *Theridula opulenta* (Walckenaer)

Linyphiidae

(Linyphiinae)
Agyneta cauta (O. P.-Cambridge)
Allomengea scopigera (Grube)
Bathyphantes anceps Kulczynski
Bathyphantes canadensis (Emerton)
Bathyphantes crosbyi (Emerton)
Bathyphantes gracilis (Blackwall)
Bathyphantes simillimus (L. Koch)
Bathyphantes pullatus (O. P.-
Cambridge)
Centromerus bicolor (Blackwall)
Centromerus sylvaticus (Blackwall)
Estrandia grandaeva (Keyserling)
Helophora insignis (Blackwall)
Lepthyphantes complicatus
(Emerton)
Lepthyphantes leprosus (Ohlert)
Lepthyphantes minutus (Blackwall)
Lepthyphantes nebulosus
(Sundevall)
Lepthyphantes nigriventris (L.
Koch)
Lepthyphantes tenuis (Blackwall)
Macrargus multesimus (O. P.-
Cambridge)
Meioneta mollis (O. P.-Cambridge)
Microlinyphia impigra (O. P.-
Cambridge)
Microneta viaria (Blackwall)
Neriene clathrata (Sundevall)
Neriene radiata (Walckenaer)
(syn. *marginata* C. L. Koch)
Neriene peltata (Wider)
Oreonetides vaginatus (Thorell)
Porrhomma convexum (Westring)

(Erigoninae)
Aulacocyba subitanea (O. P.-
Cambridge)
Caledonia evansi (O. P.-Cambridge)
Caledonia proterva (L. Koch)
Carorita limnaea (Crosby and
Bishop)
Collinsia holmgreni (Thorell)
Collinsia lapidicola (Soerensen)
Collinsia spetsbergensis (Thorell)
Collinsia thulensis (Jackson)
Cornicularia cuspidata (Blackwall)
Cornicularia karpinskii (O. P.-
Cambridge)
Diplocentria bidentata (Emerton)
Diplocephalus cristatus (Blackwall)
Dismodicus bifrons Blackwall
Entelecara media Kulczynski
Eperigone maculata (Banks)
Erigone arctica (White)
Erigone arctophylacis Crosby and
Bishop
Erigone atra (Blackwall)
Erigone dentigera (O. P.-Cambridge)
Erigone dentipalpus (Sundevall)
Erigone longipalpis (Sundevall)
Erigone psychrophila Thorell
Erigone tirolensis L. Koch
Erigone sibirica Kulczynski
Erigone whymperi O. P.-Cambridge)
Hilaira frigida (Thorell)
Hilaira glacialis (Thorell)
Hilaira herniosa (Thorell)
Hilaira laeviceps (L. Koch)
Hilaira nubigena Hull
Hilaira vexatrix (O. P.-Cambridge)
Hillhousia misera (O. P.-Cambridge)
Hybocoptus aquilonaris (L. Koch)
Hypselistes florens (O. P.-
Cambridge)
Islandiana alata (Emerton)

Maso sundevalli (Westring)
Monocephalus parasiticus (Westring)
Minyrioloides trifrons (O. P.-
Cambridge)
Ostearius melanopygius (O. P.-
Cambridge)
Pocadicnemis pumila (Blackwall)
Rhaebothorax borealis Jackson
Rhaebothorax lapponicus Holm
Rhaebothorax paetulus (O. P.-
Cambridge)
Scotinotylus antennatus (O. P.-
Cambridge)
Sisicus apertus (Holm)
Tibioplus nearcticus Chamberlin
and Ivie
Tiso aestivus (L. Koch)
Trachynella nudipalpis (Westring)
Trichopterna mengei Simon
Utopiellum mirabile (L. Koch)
Walckenaera vigilax (Blackwall)
Wideria antica (Wider)
Zornella cultrigera (L. Koch)

Tetragnathidae

Pachygnatha clercki Sundevall
Tetragnatha extensa (Linnaeus)

Araneidae

C *Argiope trifasciata* (Forskal)
Araneus diadematus Clerck
Araneus marmoreus Clerck
Araneus nordmanni (Thorell)
Araneus saevus (L. Koch)
Araniella cucurbitina (Clerck)
Araniella displicata (Hentz)
Cercidia prominens (Westring)
Cyclosa conica (Pallas)
Hyposinga pygmaea (Sundevall)
Meta menardi (Latreille)
Nuctenea cornuta (Clerck)

Nuctenea patagiata (Clerck)
Nuctenea sclopetaria (Clerck)
C *Neoscona nautica* (L. Koch)
Zygiella atrica (C. L. Koch)
Zygiella dispar (Kulczynski)
Zygiella x-notata (Clerck)

Agelenidae

Tegenaria agrestis (Walckenaer)
C *Tegenaria domestica* (Clerck)
Tegenaria pagana C. L. Koch
Tegenaria gigantea Chamberlin and
Ivie
(syn. *propinqua* Harpenden)
Tegenaria larva Simon
(syn. *praegrandis* Fox)

Hahniidae

Hahnia ononidum Simon

Lycosidae

Alopecosa aculeata (Clerck)
Alopecosa mutabilis (Kunczynski)
Arctosa alpigena (Doleschall)
Arctosa insignita (Thorell)
Pardosa palustris (Linnaeus)
Pardosa saltuaria (L. Koch)
Pardosa tesquorum (Odenwall)
Pirata insularis Emerton
Pirata piraticus (Clerck)
Trochosa terricola (Thorell)

Clubionidae

Chiricanthium mildei L. Koch
Clubiona kulczynskii Lessert
Clubiona lutescens Westring
Clubiona norvegica Strans
Clubiona pallidula (Clerck)
Clubiona trivialis C. L. Koch
Micaria pulicaria (Sundevall)

Gnaphosidae

Gnaphosa microps Holm
Gnaphosa muscorum (L. Koch)
Gnaphosa orites Chamberlin
Haplodrassus signifer (C. L. Koch)
Sosticus loricatus (L. Koch)
Scotophaeus blackwalli (Thorell)
Zelotes rusticus (L. Koch)
Zelotes subterraneus (C. L. Koch)

Sparassidae

T *Heteropoda venatoria* (Linnaeus)

Thomisidae

Misumena vatia (Clerck)
Ozyptila gertschi Kurata
Ozyptila praticola (C. L. Koch)
Ozyptila septentrionalium L. Koch
Ozyptila sincera Kulczynski
Ozyptila trux (Blackwall)
Xysticus labradorensis Keyserling
Xysticus luctuosus (Blackwall)
Xysticus obscurus Collett

Philodromidae

Philodromus alascensis Keyserling
Philodromus cespitum (Walckenaer)
Philodromus dispar (Walckenaer)
Philodromus histrio (Latreille)
Philodromus rufus Walckenaer
Thanatus arcticus Thorell
Thanatus coloradensis Keyserling
Thanatus formicinus (Clerck)
Thanatus striatus C. L. Koch
Thanatus vulgaris Simon
Tibellus maritimus (Menge)
Tibellus oblongus (Walckenaer)

Salticidae

T *Hasarius adansoni* (Audouin)
T *Menemerus bivittatus* (Dufour)
Neon reticulatus (Blackwall)
Phlegra fasciata (Hahn)
Salticus scenicus (Clerck)
Sitticus fasciger (Simon)
Sitticus finschi (L. Koch)
Sitticus lineolatus (Grube)

Most of our American spiders differ specifically and others generically from those of temperate Europe, but the relationship of the respective faunas is nevertheless clearly apparent. Broader comparison with all of Palearctica is not presently possible because of lack of information on the eastern portions of this realm. The respective mygalomorph faunas are mostly restricted to the southern parts of each area and, because of recent studies, can be described as noteworthy in terms of numbers of species and distinctive genera; only *Ummidia* is common to both. The family *Theraphosidae* is represented by many tarantula species of the genera *Dugesiella* and *Aphonopelma* in our Southwest. The typical trapdoor spiders occur in numbers in both areas; in the Mediterranean region many species of *Nemesia*, in the United States many species of *Myrmekiaphila*, *Actinoxia*, and *Aptostichus*. The derivative genus *Cyclocosmia*, with three species in our Southeast and adjacent Mexico, plugs the bottom of its burrow with its leathery abdomen. The generalized mygalomorphs are far better represented in Nearctica where two families (*Mecicobothriidae* and *Antrodiaetidae*) are almost exclusively represented. The few purse-web spiders (*Atypidae*) are of quite similar appearance in both regions, with three species in Europe and six in the eastern United States.

The cribellate families are strongly represented in the entire Holarctic region but there are notable differences in composition: the four-lunged true spiders have four species of *Hypochilus* in the United States; many *Eresidae* and *Urocteidae* are exclusive to the Mediterranean region; the multitudinous *Dictynidae* and *Amaurobiidae*, with scores of *Dictyna*, *Amaurobius*, and *Callobius*, are strongly represented in both realms; and each region has a few species of *Hyptiotes*, makers of a triangular snare. The primitive hunters and weavers also live in both realms, but again there are notable differences: the primitive eight-eyed *Plectreuridae* and their six-eyed cousins, the *Diguetidae*, are exclusively American; matching these are the numerous six-eyed *Dysderidae* remarkably developed in the Mediterranean region; the lucifugous *Leptonetidae* prosper as cave spiders in both regions and many are eyeless; and as a final vestige of kinship can be mentioned the tiny eyeless *Telema tenella* of caves in western France matched by several similar eyed and eyeless species in our northwestern states. Some of the more specialized vagrant groups are remarkably developed in the Holarctic region and these live mostly on the ground or under debris. The species of *Gnaphosa*, *Zelotes*, and *Drassodes* (all of family *Gnaphosidae*) are familiar in both areas. The species of *Clubiona* in the northern United States belong to the same groups as those of Europe, but up to the present time we note only five as being identical, and two of these are recent immigrants into the United States. The crab spiders (*Thomisidae* and *Philodromidae*) are strongly represented by the genera *Xysticus*, *Ozyptila*, and *Philodromus*, genera largely missing from the tropics. The running wolves of the genera *Lycosa*, *Alopecosa*, *Pirata*, and *Pardosa* show a parallel growth in both regions. A surprisingly small number of these vagrants have been shown to be the same all over the Holarctic realm.

It is chiefly in the sedentary aerial spiders, most of which are well known for their ballooning talents, that one finds many American spiders identical with those from Eurasia; this is well shown by the list of *Theridiidae*, *Linyphiidae*, and *Argiopidae* common to both areas. If one were to name the single group largely typical of the temperate zone it would be the *Linyphiidae*. This great family, which includes the multitudinous tiny dwarf spiders that have modified their heads in singular ways (*Erigoninae*) and the somewhat larger sheet spinners (*Linyphiinae*), has had an unparalleled development in the Holarctic region. It makes up a substantial percentage of the fauna of Canada and most of our states: about 30 percent in New York State, 47 percent in Alaska, similarly high percentages in the woods and tundra of Canada, and 54 percent in Greenland. The representation decreases toward the South, to about 16 percent in Georgia, with even smaller percentages in our southwestern states. In Europe the *Linyphiidae* are similarly abundant, with 29 percent in France, 43 percent in the British Isles, and 41 percent in Norway and Sweden. In the tropics these spiders are largely replaced by other types; in Brazil only 1.2 percent of the total fauna belongs to the *Linyphiidae*.

The differences between the faunas of Europe and the United States are largely due to those species, genera, and larger groups that probably have been derived from the South, or represent a remnant of the subtropical fauna that once occupied a more substantial portion of the temparate zone. In the United States most of them occur in the southeastern states, in the arid southwestern part of our country (making up what is often called a Sonoran fauna), and also in the mild to severe Californian zone. In Europe an analogous fauna exists in the Mediterranean subregion of southern Europe and North Africa. In both areas are found many trap-door spiders, large lycosids, pisaurids, and ctenids, and representatives of tropical genera (including the mobile tropicopolitan species) that have reached their northern limits of distribution. A very few major groups are represented in only one region: the families *Dysderidae, Argyronetidae, Urocteidae*, and *Eresidae* have no endemics in Nearctica: whereas the families *Mecicobothriidae, Antrodiaetidae, Hypochilidae, Plectreuridae, Diguetidae, Homalonychidae*, and *Caponiidae* are almost exclusively North American.

The following list of taxa shows graphically the similarities and differences in the spider faunas of the Nearctic and Palearctic regions. The figures for the latter are taken mostly from Eduard Reimoser's checklist of the Palearctic spiders. Those from the American column are derived from various catalogs, but also supplemented by unpublished information. All larger numbers from either region could appropriately be preceded by a plus-or-minus sign, since they are for the most part *approximations* of the true situation.

NUMBER OF SPECIES

	American		European
Atypidae	6		3
Mecicobothriidae	7		0
Antrodiaetidae	25		0
Theraphosidae	40		1
Cyclocosmia	3		0
Hypochilidae	4		0
Dysderidae	1	(cosmopolitan)	50
Caponiidae	4		0
Plectreuridae	30		0
Diguetidae	5		0
Leptonetidae	45		30

NUMBER OF SPECIES (*continued*)

	American	European
Telemidae	5	1
Pholcidae	30	20
Loxoscelidae	10	1
Eresidae	0	26
Urocteidae	0	3
Zodariidae	3	54
Argyronetidae	0	1
Anyphaenidae	45	8
Zelotes (*Zelotes*)	26	118
Zelotes (*Drassyllus*)	75	10
Heliophanus	0	56
Euophrys	2	45
Phidippus	55	0
Metaphidippus	55	0
Pellenes (*Habronattus*)	50	0

In summary, then, it can be said that the rich and varied North American spider fauna is similar to and for the most part derived from the same sources as the temperate Eurasian fauna. A modest segment comes from the American tropics, and its present distribution is largely limited to the extreme southern and southwestern states. A few archaic types still persist within our boundaries: four species of *Hypochilus* in three disjunct centers from Appalachia to California; numerous species of *Plectreuridae*, *Diguetidae*, and *Caponiidae* mainly in our southwestern states and California; *Mecicobothriidae* in Arizona and the Californian region; and three genera of *Antrodiaetidae* from Appalachia across many of our states. Although the general character of the North American fauna in now known, many of the details are still vague, and can be clarified by more revisional studies and further exploration of neglected regions.

The following arrangement of spider *families* and higher systematic divisions is one of convenience adopted as a practical expedient for use in this volume. All the families listed have representatives in the North American fauna.

SUBORDER MYGALOMORPHAE

Atypoidea
(*The Atypical Mygalomorphs*)

Mecicobothriidae
Antrodiaetidae
Atypidae

Ctenizoidea
(*The Typical Mygalomorphs*)

Dipluridae
Theraphosidae
Ctenizidae

SUBORDER ARANEOMORPHAE

Hypochiloidea
(The Four-Lunged True Spiders)

Hypochilidae

Filistatoidea
(The Filistatids)
Filistatidae

Dictynoidea
(The Typical Cribellate Spiders)
Amaurobiidae
Tengellidae
Acanthoctenidae
Dictynidae
Uloboridae
Dinopidae
Oecobiidae

Araneoidea
(The Aerial Web Spinners)

Mimetidae
Nesticidae
Linyphiidae
Anapidae
Symphytognathidae
Mysmenidae
Theridiidae
Tetragnathidae
Theridiosomatidae
Araneidae

Lycosoidea
(The Hunting Spiders)

Hersiliidae
Agelenidae
Hahniidae
Oxyopidae
Pisauridae
Lycosidae

Plectruroidea
(The Primitive Hunters and Weavers)

Plectreuridae
Diguetidae
Loxoscelidae
Scytodidae
Dysderidae
Segestriidae
Caponiidae
Oonopidae
Pholcidae
Leptonetidae
Telemidae
Ochyroceratidae

Clubionoidea
(The Running Spiders)

Prodidomidae
Zodariidae
Homalonychidae
Anyphaenidae
Clubionidae
Gnaphosidae
Ctenidae
Selenopidae
Sparassidae
Philodromidae
Thomisidae
Salticidae

Glossary

(Definitions with especial reference to spiders)

Abdomen. The posterior division of the spider body, comprising the pedicel and usually largely unsegmented saclike portion bearing the spinnerets.

Anal tubercle. The small caudal tubercle bearing the anal opening; the post-abdomen.

Antennae. The segmented sensory organs often termed "feelers," borne on the heads of insects, crustacea, etc., but missing in all arachnids.

Appendages. Parts or organs (such as legs, spinnerets, chelicerae) that are attached to the body.

Arachnida. A principal division, or class, of the air-breathing arthropods, the arachnids, including the scorpions, mites, spiders, etc.

Arachnologist. One who studies the arachnids.

Araneae. The ordinal name of all spiders; same as Araneida.

Araneology. The branch of zoology that treats only of spiders.

Arthropod. The jointed-legged animals, such as centipedes, millipedes, insects, crustaceans, spiders, scorpions, and many other less well-known types; the members of the phylum Arthropoda.

Atriobursal orifice. The opening of seminal receptacle of female spider.

Attachment disk. The series of tiny lines that serve to anchor the draglines of spiders.

Autophagy. The eating of an appendage shed from the body by autotomy or otherwise.

Autospasy. The loss of appendages by breaking them at a predetermined locus of weakness when pulled by an outside force; frequent in spiders and arachnids.

Autotomy. The act of reflex self-mutilation by dropping appendages; unknown in the arachnids.

Ballooning. Flying through the air on silken lines spun by spiders.

Book lungs. The respiratory pouches of the arachnids, filled with closely packed sheets or folds to provide maximum surface for aeration; believed to be modified, insunk gills.

Calamistrum. The more or less extensive row of curved hairs on the hind metatarsi, used to comb the silk from the cribellum.

Carapace. The hard dorsal covering of the cephalothorax in the Arachnida.

Cephalothorax. The united head and thorax of Arachnida and Crustacea.

Chelicerae. The pincerlike first pair of appendages of the arachnids; in spiders two-segmented, the distal portion or fang used to inject venom from enclosed glands into the prey.

Chorion. The outer covering or shell of the spider or insect egg.

Claw tufts. The pair of tufts of adhesive hairs present below the paired claws at the tip of the tarsi of many spiders.

Colulus. The slender or pointed appendage immediately in front of the spinnerets in some spiders; in others greatly reduced or seemingly missing; the homologue of the anterior median spinnerets or cribellum.

Coxa. The basal segment of the leg by means of which it is articulated to the body.

Coxal glands. The excretory organs of arachnids, in spiders located opposite the coxae of the first and third legs, that collect wastes into a saccule and discharge them through tubes opening behind the coxae; homologous with the nephridia of Peripatus, etc.

Cribellum. A sievelike, transverse plate, usually divided by a delicate keel into two equal parts, located in front of the spinnerets of many spiders; the modified anterior median spinnerets.

Cuticle. The hard outer covering of an arthropod.

Deutovum. The resting, spiderlike stage following the shedding of the chorion of the egg; the second egg.

Dorsum. In general, the upper surface.

Ecdysis. The process of casting the skin; molting.

Embolus. The intromittent portion of the male copulatory organ, containing portion of the ejaculatory duct.

Endite. The plate borne by the coxa of the pedipalps of most spiders, used to crush the prey; the maxilla.

Epigynum. The more or less complicated apparatus for storing the spermatozoa, immediately in front of the opening of the internal reproductive organs of female spiders.

Femur. The thigh; usually the stoutest segment of the spider's leg, articulated to the body through the trochanter and coxa and bearing the patella and remaining leg segments at its distal end.

Genitalia. The sclerotized genital structures of male and female spiders.

Gonopore. The reproductive orifice of the female.

Hackled band. The composite threads of the cribellate spiders, spun by cribellum and combed by the calamistrum.

Instar. The period or stage between molts in the postembryonic development of arthropods.

Integument. The outer covering or cuticle of the spider or insect body.

Labium. The lower lip of spiders forming the floor of the mouth cavity.

Maxillae. In spiders, used as a synonym of the endites or coxae of the pedipalps.

Metatarsus. A principal segment of the legs, the sixth from the base, with tibia at its base and tarsus at its apex.

Molting. The periodic process of loosening and discarding the cuticle, accompanied by the formation of a new cuticula.

Mygalomorph. The members of the suborder Mygalomorphae, the tarantulas, trap-door spiders, and all their kind.

Nephridia. Tubular structures used as excretory organs in annelids, mollusks, etc.

Ocelli. The simple eyes of insects.

Ostia. The slitlike openings into the heart of spiders and insects.

Palpus. The segmented appendage of the pedipalp, exclusive of coxa and endite; in female spiders, simple; in males, bearing a reproductive organ.

Patella. A segment of the leg between the femur and tibia in the arachnids.

Pedicel. The attenuated first abdominal segment, or waist, of spiders, which joins the abdomen to the cephalothorax.

Pedipalps or Pedipalpi. The second pair of appendages of the head of spiders, consisting of a coxal portion to aid in crushing prey and a distal appendage or palpus.

Pheromone. The chemical substances on the substratum, on the bodies, or possibly in the air, detectable by male and female arachnids, chiefly during courtship and mating.

Postabdomen. In spiders, the anal tubercle; the fused vestigial segments of the abdomen.

Receptors. The sense organs; specialized structures of the integument that respond to external stimuli.

Scales. Flattened, modified setae of spiders.

Sclerotized. Hardened by deposition of sclerotin or other substances in the cuticule.

Scopula. A small, dense tuft or more extensive brush of hairs or setae.

Segment. A ring, somite, or subdivision of the body or of an appendage between areas of flexibility.

Setae. The slender hairlike or spinelike appendages of the body.

Sexual dimorphism. A difference in form, color, size, etc., between sexes of the same species.

Sigilla. The impressed, suboval, clear areas on the sternum of some spiders.

Spermathecae. The vessels or receptacles in the epigyna of female spiders that store the spermatozoa of the males.

Spermatophore. The sclerotized vehicle that carries the sperm mass of arachnids and some other animals.

Spermatozoa. The mature sperm cells.

Sperm induction. The process of transferring the spermatozoa from the genital orifice beneath the base of the abdomen into the receptacle in the male palpus.

Sperm web. A web of few or many threads on which male spiders deposit the semen preparatory to taking it into the palpus.

Spiderling. A tiny, immature spider, usually the form just emerged from the egg sac.

Spinnerets. The fingerlike abdominal appendages of spiders through which the silk is spun.

Spiracle. A breathing pore or orifice leading to tracheae or book lungs.

Stadium. The interval between the molts of arthropods; instar; a period in the development of an arthropod.

Sternum. A sclerotized plate between the coxae marking the floor of the cephalothorax.

Tarsus. The foot; the most distal segment of the legs, which bears the claws at its tip.

Tergites. Dorsal sclerites on the body; the hard plates on the abdomen of the atypical tarantulas that indicate the segmentation.

Thorax. The second region of the body of insects that bears the legs; in spiders, fused with the head to form the cephalothorax.

Tibia. The fifth division of the spider leg, between the patella and metatarsus.

Tracheae. The air tubes in insects; in spiders, tubular respiratory organs of different origin, thought to be modified book lungs.

Zygote. The fertilized egg.

Bibliography

The following references will be found useful to those readers who wish to delve more deeply. For the most part, they are books of general interest, likely to be available in larger libraries; many are furnished with excellent bibliographies. Some journal references make possible finding of quotations in the text and permit further reading of the special subjects. Advanced students can find a complete list of spider literature current to 1939, systematically classified by subject, in the first volume of the *Bibliographia Araneorum* of Pierre Bonnet.

Berland, L., *Les Arachnides, Encyclopedie entomologique*, XVI. Paris: Paul Lechevalier & Fils, 1933.

Bonnet, P., *Bibliographia Araneaorum*, vol. 1. Toulouse: (publ. by author), 1945.

Bristowe, W. S., *The Comity of Spiders*, vols. 1 and 2. London: The Ray Society, 1939 and 1941.

Bristowe, W. S., The Courtship of British Lycosid Spiders, and its probable Significance. *Proc. Zool. Soc. London* 2: 317-347 (1926).

Bristowe, W. S., *The World of Spiders, The New Naturalist.* London: Collins Clear-Type Press, 1958.

Comstock, J. H., *The Spider Book.* New York: Doubleday, 1912; revised edition, 1940, by W. J. Gertsch.

Emerton, J. H., *The Common Spiders of the United States.* Boston: Ginn, 1902.

Fabre, J. H., *The Life of the Spider.* New York: Dodd, Mead, 1913.

Forster, R. R. and Forster, L. M., *New Zealand Spiders, An Introduction.* Auckland and London: Collins Brothers & Co., 1973.

Kaston, B. J., The Senses Involved in the Courtship of the Vagabond Spiders. *Entomologia Americana* 16: 97-167 (1936).

Kaston, B. J., *How to Know the Spiders*, 3rd ed. Dubuque: Wm. C. Brown Co., 1978.

Levi, H. W. and Levi, L. R., *Spiders and Their Kin, A Golden Nature Guide.* New York: Golden Press, 1968.

Main, B. Y., *Spiders, The Australian Naturalist Library.* Sydney and London: William Collins Publishers, 1976.

McCook, H. C., *American Spiders and Their Spinningwork*, vols. 1–3. Philadelphia: (publ. by author), 1889–1894.

McKeown, K. C., *Spider Wonders of Australia.* Sydney: Angus & Robertson, 1936.

Moggridge, J. T., *Harvesting Ants and Trap-Door Spiders.* London: L. Reeve & Co., 1873.

Montgomery, T. H., Studies on the Habits of Spiders, Particularly Those of the Mating Period. *Proc. Acad. Nat. Sci. Philadelphia* **55**: 59–119 (1903).

Montgomery, T. H., The Significance of the Courtship and Secondary Sexual Characters of Araneids. *The American Naturalist* **44**: 151–177 (1910).

Nielsen, E., *The Biology of Spiders*, vol. 1 (in English), vol. 2 (in Danish). Copenhagen: Levin & Munksgaard, 1932.

Peckham, G. W. and Peckham, E. G., Observations on Sexual Selection in Spiders of the Family Attidae. *Occasional Papers Nat. Hist. Soc., Wisconsin* **1**: 1–60 (1889).

Peckham, G. W. and Peckham, E. G., Additional Observations on Sexual Selection of the Family Attidae. *Occasional Papers Nat. Hist. Soc., Wisconsin* **1**: 117–151 (1890).

Peckham, G. W. and Peckham, E. G., Revision of the Attidae of North America. *Trans. Wisconsin Acad. Sci.* **16**: 355–646 (1909).

Petrunkevitch, A., Sense of Sight, Courtship and Mating in *Dugesiella hentzi* (Girard), a Theraphosid Spider from Texas. *Zool. Jahrbucher Syst.* **31**: 355–376 (1911).

Petrunkevitch, A., Tarantula versus Tarantula-Hawk: A Study in Instinct. *J. Exp. Zool.* **45**: 367–394 (1926).

Savory, T. H., *The Biology of Spiders.* London: Sidgwick & Jackson, 1928.

Savory, T. H., *The Arachnida.* London: E. Arnold, 1935.

Savory, T. H., *The Spider's Web, The Wayside and Woodland Series.* London and New York: Frederick Warne & Co., 1952.

Schwarz, H. F., Spider Myths of the American Indians. *Natural History, Journal of the American Museum of Natural History* **21**: 382–385 (1921).

Thorp, R. W. and Woodson, W. D., *Black Widow.* Chapel Hill: University of North Carolina Press, 1945.

Warburton, C., *Spiders.* Cambridge: The University Press, 1912.

Index